Invention
-in-
America

On this 19th-century ad promoting inventions of a rocking horse and "self-propelling ice sleighs," the small print advised purchasers to apply the motion-making attachments themselves.

Invention
-in-
America

Russell Bourne

Fulcrum Publishing
Golden, Colorado

Published in cooperation with the Library of Congress

Jacket photos: *Front*—Edison's light bulb, first flight at Kitty Hawk. *Back* (left to right)—
Lynn mill workers, by Frances B. Johnston, 1895; domestic sewing machine, lithography by
W. J. Morgan & Co.; Hamlin's Wizard Oil, Calvert Litho Co.; Wolcott's Instant Pain
Annihilator, 1863 lithograph; Universal Food Chopper, 1899 lithograph.

LIBRARY OF CONGRESS CATALOGING-IN-PUBLICATION DATA
Bourne, Russell.
    Invention in America / Russell Bourne.
        p.   cm.
    Includes bibliographical references and index.
    ISBN 1-55591-231-1 (hardcover : Alk. paper)
    1. Inventions—United States—History. 2. Technology—Social aspects—United
States—History. I. Title.
T21.B693   1995
609.73—dc20   95–35784
                                                                                  CIP
Printed in the United States of America

0 9 8 7 6 5 4 3 2 1

Fulcrum Publishing
350 Indiana Street, Suite 350
Golden, Colorado 80401-5093
800/992-2908

Dedicated to D. F. B.

CONTENTS

MECHANICS' MAGAZINE,
Museum, Register, Journal and Gazette.

"The most valuable gift which the Hand of Science has ever
yet offered to the Artisan." *Dr. Birkbeck.*

No. 1. Am. Ed.]   SATURDAY, FEBRUARY 5, 1825.   [Price $4 per ann.

"They helped every one his neighbour, and every one said to his brother, Be of good
courage. So the carpenter encouraged the goldsmith, and he that smootheth with the
hammer him that smote the anvil, saying, It is ready for the soldering; and he fastened
it with nails, that it should not be moved."—*Isaiah* xli. 6, 7.

NEW PIT-SAW.

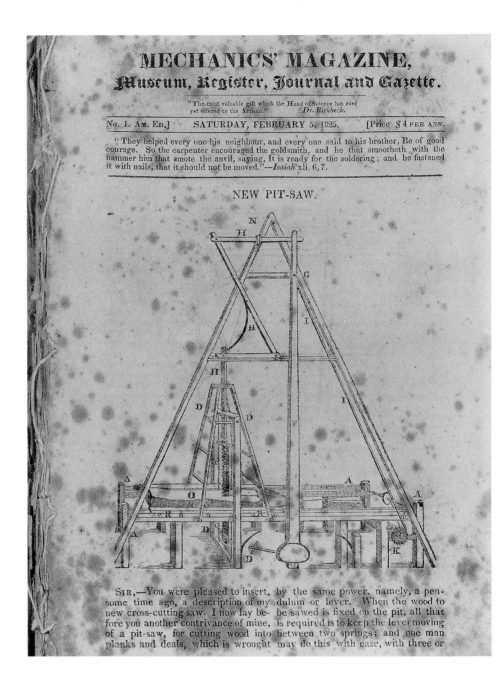

Sir,—You were pleased to insert, some time ago, a description of my new cross-cutting saw. I now lay before you another contrivance of mine, of a pit-saw, for cutting wood into planks and deals, which is wrought by the same power, namely, a pendulum or lever. When the wood to be sawed is fixed on the pit, all that is required is to keep the lever moving between two springs; and one man may do this with ease, with three or

FOREWORD

Here in the seventh "Library of Congress Classics" volume, *Invention in America,* historian Russell Bourne draws on the unparalleled collections of the world's largest library to enhance his chronicle of key American inventions and the varying fortunes of their creators.

One great American inventor, in the grandest sense of that term, was Thomas Jefferson—lawyer, architect, musician, political philosopher, linguist, and scientist. As the principal founder of the Library of Congress, Jefferson believed that there was no subject "to which a member of Congress may not have occasion to refer." This approach, endorsed by the Congress, has guided the Library ever since, as it has assembled collections of matchless breadth and depth, which now total more than 107 million items. Thus, as Mr. Bourne broadly illuminates the conditions that fostered technological advance in the United States, his narrative reflects the magnificent historical record preserved in the Library of Congress.

Special collections, such as the papers of Wilbur and Orville Wright and Alexander Graham Bell, document the trials, errors, and triumphs of those pioneers, while one entire division of the Library—that of Motion Picture, Broadcasting, and Recorded Sound—is a showcase for the early inventions of Thomas Edison and others.

The Library's prints and photographs, used here, superbly document not only the technical history of invention but its social history as well, for as Mr. Bourne makes clear, no less important than the invention are the patterns of its adoption and use. And in the diversity of other Library materials ranging from periodicals, legal documents, music, children's books, and advertising art to architectural drawings, comics, and maps, one can also view all aspects of technological advance; a new device can be seen making the progression from innovation to everyday companion and part of the national landscape.

Those who have read Russell Bourne's previous volume in this series, the lively study of American transportation entitled *Americans on the Move,* will recognize his style here; thoughtfulness, freshness of approach, highly readable prose, and an eye for the significant details.

It is our hope that series such as the "Library of Congress Classics" will prompt readers to explore their own local libraries, and to appreciate the rich intellectual heritage that the Library of Congress preserves for the nation.

—James H. Billington
The Librarian of Congress

Invention
-*in*-
America

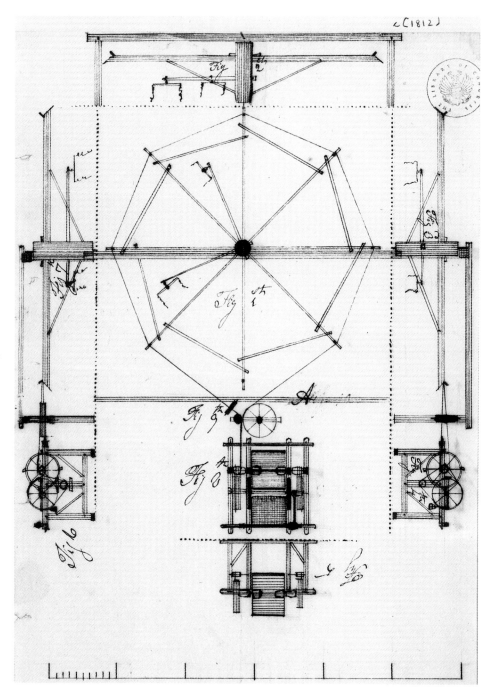

In his 1812 diagram of a spinning jenny, a complex frame for spinning yarn, Thomas Jefferson analyzed this generative invention of the 19th-century Industrial Revolution.

ONE

THE DYNAMICS OF AMERICAN INVENTION

his old ax. If it were on display in a museum, it would probably not catch your eye or cause you to slow down for a closer look. But this first-rate ax of the early American frontiersman, made of iron head and wooden shaft, not unlike the ax we use today, should be viewed with considerable awe. It worked superbly well compared with other technologies intended for the same purpose, allowing American colonials to fell three times as many trees within the same period as their European cousins. This unique tool also expressed the ancient Greek ideal of perfection, the identity of utility and beauty. Furthermore, it demonstrated in its own time and still demonstrates today the dynamics of American invention, dynamics which would drive our democratic, capitalistic, industrialized culture from the seventeenth into the twentieth century.

The secret of this particular invention is that the back part of the ax head, the poll, perfectly balances the front part, the cutting edge. That makes the ax much easier to swing up over the shoulder and back down into the tree. Additionally, the curved, wooden handle is designed to be just the right length for the individual axman; he can keep working for hours at a time at a natural rhythm, maintaining his own stance. Thanks to his easy-swinging ax, the historic American frontiersman could clear and plant as many as ten acres a year—driving back the wilderness and claiming the continent. If, as historians assert, the saddle and stirrup enabled French knights of the Middle Ages to conquer Europe, it might be said that the balanced ax, wielded by English- and German-speaking settlers, was the decisive tool in the settlement of agricultural North America.

Along came Eli Whitney's cotton gin, Samuel Colt's revolver, Alexander Graham Bell's telephone—these and many other uniquely American inventions, all key tools for what is known as Progress. How Americans zigzagged through successive historic eras, from the seventeenth century age of wood and iron to the early twentieth century age of electricity and petroleum, can best be comprehended by keeping an eye on those tools, results of the distinctive dynamics of American invention. The most obvious three dynamics may be identified as follows:

The experimental procedures of Thomas Jefferson (seen in engraving below) involved scientific observation and note-taking—plus some subjectivity. In his notes at right on a possible cure for "head ach," he chronicles patients on whom the calomel had a desired effect. But in the note at the bottom, he expresses doubt that the sickness of a "negro" patient was genuine.

[1792-1796]

Recipe for the head-ach called the Sun-pain

6. grains of Calomel taken at night, without any thing to work it off, or any other matter whatever.

Mr. Willis of Georgia has had this head-ach to a dangerous extremity, & tried bark ineffectually. a drunken Doctr. recommended the above dose of Calomel, it relieved him from the next fit. he had it some years after, & the same medicine relieved him again as quickly.

he has cured his son of it also, & was obliged to give it twice at an interval of 2. or 3 days.

he communicated it to mrs Broadnax a sister of Colo. Fields Lewis who had the same species of headach. it cured her.

she has communicated it to others who have tried it with never-failing success.

Apr. 10. 1792.

Sep. 1. 96. a negro man had been 3 weeks laid up with the periodical headach or eyeach. I gave him a dose of salts in the morning & in the night 3. drachms of bark. recollecting then the above prescription I gave no more bark, but the next night gave 6 grs. of calomel. he missed his fit the next day. but note. I was not perfectly satisfied that his sickness was genuine.

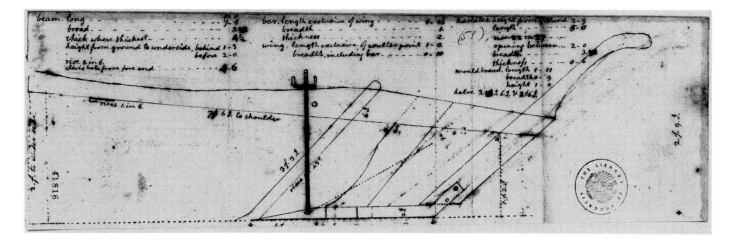

1. A successful invention embodied the skills of many people, usually superior mechanics, working for themselves or for an employer—a combination of talents, plus fortunate timing.

2. A successful invention, designed to serve a certain purpose, was usually accompanied by a massive promotional program; indeed, the acceptance of the new technology seemed to be in direct proportion to the success of that sales campaign.

3. A successful invention, which usually required intensive capital-raising efforts, ultimately resulted in a major industrial advance; indeed, the invention might be seen as but one basic part of a coordinated, capitalistic push into a new industrial epoch.

Jefferson's detailed drawing for a mold-board plow, with precise measurements for all the working parts, resulted from intensive experiments in the field—the brilliance of his mind collaborating with the pragmatic ideas and applied muscle-power of his black workers.

Thomas Jefferson, who sired so many inventions and set up the inventor-protective U.S. Patent Office (not to mention his role in the creation of the Library of Congress, whose archives record these inventions), summed up the process described in the first dynamic. "One idea leads to another," he explained, and "that to a third, and so on through a course of time until someone, with whom no one of these is original, combines all together, and produces what is justly called a new creation."

By all three counts, the pivotal American inventions encountered in this and the following chapters tend to have been the work of not one genius but of several inspired mechanics, they tend to have been extraordinarily and self-consciously promoted, and they were most often parts of larger systems involving heavy and competitive capital. This extended definition of "invention" encompasses only America's material technologies—tools for the hand—not literary inventions or inventions of the mind (like the most powerful and magnificent invention in American History, the U.S. Constitution). Certain culturally inclined observers of the changing American scene (such as Harvard's late, great Perry Miller) take the other view, focusing on the unworldly power of the human mind and spirit to shape a culture. But how is that shaping done? This book moves in company with historians and sociologists who have chosen to look at the palpable weapons, the tangible factors which have been responsible for the mixed victories of this very materialistic and, yes, occasionally, spiritual people.

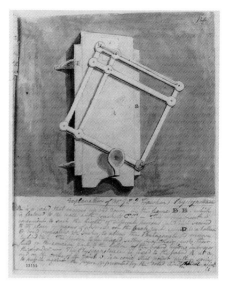

For students of the human condition, Jefferson explained the workings of John Hawkin's inventive "physiognotrace," sketched above, whose purpose was to chart accurately the oddities of a person's face and thus to figure out his character. For Monticello, he devised a hot-air heating system, shown in his drawing below right, which sent heat from a sealed, central oven to various rooms by pipes; oxygen was brought in from outside by another pipe.

Americans have reason to ponder how these historic, determining creations came into being, in such numbers, and with so great an effect. Some of these tools have been the work of unknown craftsmen such as the blacksmith who first developed the balanced ax; others have been the work of such brilliant theoreticians as Jefferson himself (who gave us the mold-board plow) or of such methodical tinkerers as Thomas Edison (who, as well as contributing to the realization of the light bulb, invented the actual business of invention). From the very beginning of American history, we have chosen to honor our inventors, to credit their better mousetraps with having, "during the course of time," turned around our economy and our society.

We have imagined ourselves to be an ever-advancing people, released from the limitations of the past by means of consecutive inventions. We have believed that the clever solution to any difficulty must always be at hand—the revolver that will win the West, the sewing machine that will liberate the woman—yet sometimes this occurs with unexpected results. And so the inventor has been hailed as hero or even god, even when his genius causes major social disruptions. This belief of ours goes way back.

Jonathan Edwards, the influential preacher of the 1730s whose fiery sermons sparked New England's "Great Awakening," might seem to have been far more interested in theology than technology. But he, too, treasured tools and inventions. Among his other thunderings and cries to heaven, Edwards called for new "contrivances" that would multiply the "efficacy of Human effort." He saw the point: Man-muscle was simply not powerful enough (even when that man was equipped with a balanced ax) to advance contemporary society beyond its ancient, static condition. While Edwards had in mind a technology that would help move his follow religionists into a "Paradise on Earth," others of his day were interested in the economic potential of waterwheels, of engines, of explosives—anything that might give struggling settlers the additional power to make their environments more productive, their lives less wearying.

This desire for better tools was, by Jonathan Edwards's time, nearly a century old, already an established cultural characteristic. In myth, the antiquity of that American

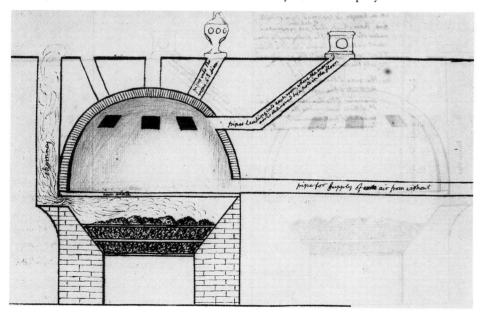

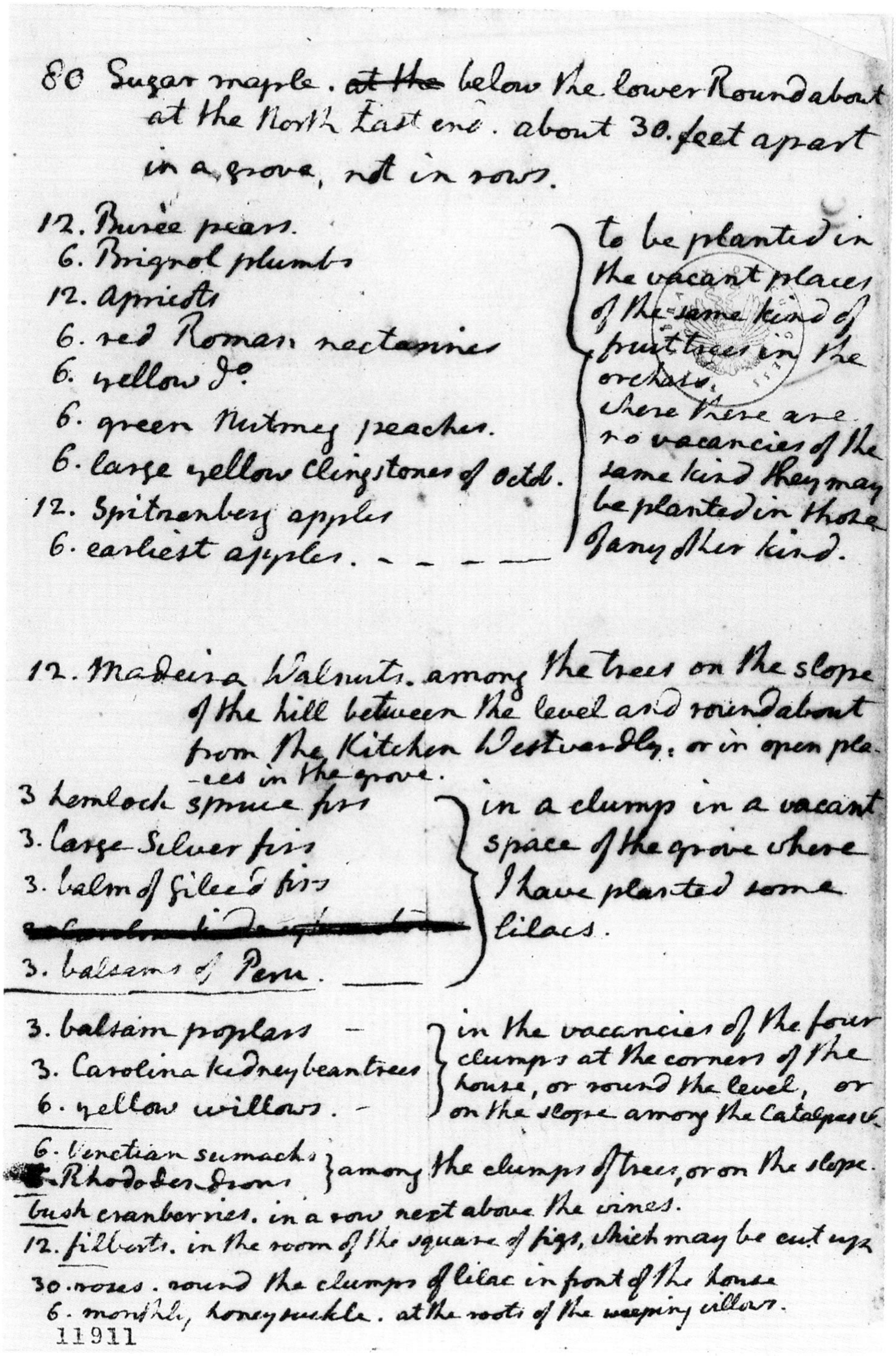

80 Sugar maple. ~~at the~~ below the lower Roundabout at the North East end. about 30. feet apart in a grove, not in rows.

12. Burée pears.
6. Brignol plumbs
12. apricots
6. red Roman nectarines
6. yellow do.
6. green Nutmeg peaches.
6. large yellow clingstones of Octob.
12. Spitsenberg apples
6. earliest apples. — — — —

to be planted in the vacant places of the same kind of fruit trees in the orchard. Where there are no vacancies of the same kind they may be planted in those of any other kind.

12. Madeira Walnuts. among the trees on the slope of the hill between the level and roundabout from the Kitchen Westwardly, or in open places in the grove.

3. hemlock spruce firs
3. large Silver firs
3. balm of Gilead firs
~~Carolina kidney~~
3. balsams of Peru.

in a clump in a vacant space of the grove where I have planted some lilacs.

3. balsam poplars —
3. Carolina kidney bean trees
6. yellow willows. —

in the vacancies of the four clumps at the corners of the house, or round the level, or on the slope among the Catalpas &c.

6. Venetian sumachs.
~~6.~~ Rhododendrons } among the clumps of trees, or on the slope.
bush cranberries. in a row next above the vines.
12. filberts. in the room of the square of figs, which may be cut up.
30. roses. round the clumps of lilac in front of the house
6. monthly honeysuckle. at the roots of the weeping willows.

11911

Horticulture and landscaping were also arts that Jefferson nurtured with science and ingenuity. At left are his exacting directions for plantings at Monticello—nothing in rows.

yearning is represented by the constantly debated story of the Pilgrims' adaptation of the Algonquin Indians' trick of placing a decaying fish in each hill of corn to fertilize the plant. The boosting of agriculture and of the general economy by invention would go on to become, half a century later, the expressed purpose of America's post-Revolution leadership: George Washington, in his famous First Address to the Congress, urged that citizens contribute "exertions of skill and genius" to the nation's growth. Only by such inventions, he believed, would the newly born United States of America be freed from servile dependence on the world's major powers (i.e., England and France).

His purposes were, of course, fulfilled. America, more than any other nation, won its freedom in war and peace and has defined itself by its inventions; American generations identify themselves by whether they, in their own time, were the ones who experienced the birth of the airplane or of the refrigerator or of the computer. For Americans to be ingenious seemed to be not only our destiny but also a byproduct of certain political choices we had made. The brilliant British scientist and theologian, Joseph Priestley, who had come here from his homeland in 1794 because of his treasonably democratic views, foresaw what would happen in his new land. "It will soon appear," he wrote, "that Republican governments, in which every obstruction is removed to the exertion of all kinds of talent, will be far more favorable to science and the arts [i.e., technology] than any monarchical government has ever been." And that keenest of all observers of early America, Alexis de Tocqueville, when he visited here during the Jacksonian era, saw this link between democracy and technology as an essential part of American culture . In an important paragraph in his four-volume *Democracy in America,* he stated:

> You may be sure that the more democratic, enlightened, and free a nation
> is, the greater will be the number of these interested promoters of scientific
> genius, and the more will discoveries immediately applicable to productive in-
> dustry confer on their authors gain, fame, and even power.

Deducing that inventiveness was intimately related to the new industrialization of our new republic, he entitled one entire chapter in his work, "What Causes Almost All Americans to Follow Industrial Callings." De Tocqueville saw the average American as too impatient to put up with the perpetual poverty of farm life; he described how, on these shores, the country person "sells his plot of ground, leaves his dwelling, and embarks on some hazardous but lucrative calling."

Yet, for all de Tocqueville's admiration for the ambitious characteristics of the American people, he regretted that their materialism, "the love of well-being," had become "the predominant taste of the nation." Furthermore, he feared that, since the Americans' technological advancements and industrial progress had removed them from the solidity of an agricultural economic base, they would constantly be subject to disastrous "economic panics."

De Tocqueville was quite accurate in reporting the trends and the dangers of his time, and in projecting what might happen thereafter. But perhaps he did not probe far enough into young America's preindustrial history to understand the deep roots of our materialistic, tool-using tendencies. He did not appreciate what this book seeks to set forth: the inherent dynamics of American inventiveness—which may or may not lead to disaster.

BEN FRANKLIN AND
THE ROOTS OF INVENTIVENESS

Long before the Jacksonian era, the era when de Tocqueville observed certain "scientific geniuses" and inventors working on our steam engines and our mass-produced firearms, a scientist named John Winthrop sought to set America on an industrial course. Son of the famed governor of the Puritan-founded Massachusetts Bay Colony, Winthrop the Younger had attended Trinity College, Dublin, studied law and served in Great Britain's European campaigns before coming to join his family in America in 1631. By 1659 he had become governor of Connecticut, and in 1664 he succeeded in merging the New Haven Colony with the rest of Connecticut under a charter from the Stuart kings.

A notable collector of books on scientific matters and a sometime practitioner of medicine, Winthrop won sufficient respect from his contemporaries on both sides of the Atlantic to be named one of the first fellows of Britain's Royal Society. Multi-talented, he made it his business to probe the nature of finance and became known as "the most sought-after man in New England." Like other inventors, he attempted many experiments which led nowhere: he tried in vain to find a method of distilling salt from sea water (a process not figured out until two centuries later when fishing communities on Cape Cod perfected the production of salt via a series of evaporation pans). But in his knowledge of and prospectings for useful metals, he was triumphant. His discovery of graphite near Sturbridge, Massachusetts, gave an early boost to the colonies' nascent pencil industry. And his successful searchings for reliable iron sources led, in 1647, to the establishment of America's first ironworks. Built above the Saugus River ten miles north of Boston, his Hammersmith Works (for

John Winthrop's ideas for distilling salt from ocean waters by a series of evaporation pans undoubtedly influenced these later salt works near the mouth of the Merrimack River.

In colonial America's age of wood and iron, mechanics developed technologies to magnify energy that was available in human and animal power as well as in water and wind. The lofty windmill photographed above was built in the 1740s on Gardiner's Island off the eastern tip of Long Island; the cut-away view at right, showing wheels for grinding grain on the mill's second floor, was prepared by the Historic American Engineering Record (HAER), whose archives are maintained by the Prints and Photographs Division of the Library of Congress.

whose funding Winthrop had to sail repeatedly back and forth to England) was a seeming industrial miracle for its day.

It produced serviceable bar iron for three decades—the only such manufacturing facility in North America—reaching a size and capacity that even today seem wondrous. Indeed, the very magnitude of the enterprise (employing fifty or sixty people laboring at three furnaces and four waterwheels) seemed a moral problem to the religious leaders of the Bay Colony. They were at that point disinclined to favor any enterprises that rivaled the church in scale and significance. These conservative Puritans were joined in their condemnation of Saugus by a new class of Boston merchants jealous of such a powerful business within their territory. They saw to it that, even though the Hammersmith Works turned out inexpensive nails for the great benefit of colonial homebuilders, it be given no protection against the under-priced ironware imported from England.

So it happened that all the elaborate industrial constructions at Saugus, the interactive complex that was probably Winthrop's greatest invention—the ingenious, multilevel sluice ways for the waterwheels, the clever separating of iron from swampy peat by means of a calcium-rich "flux" with which it was mixed, the innovative production of a liquid iron from which wrought iron bars were forged—all this clanging and hissing machinery came to a halt. Winthrop himself died in the very year when the Hammersmith Works was shut down, disappointed that, for all his success in political and business affairs, his intuitive effort to advance New England into what came to be called the Industrial Revolution had been frustrated. Yet, it must be admitted, Massachusetts bog iron—the basic substance for this industry—was inferior stuff; Hammersmith's hardware and other manufactured tools suffered from that fault. This colonial industry could only go a limited distance toward that distant revolution.

John Winthrop possessed an extraordinarily promethean mind; he'd been shrewd enough to recognize the third principle of American invention, namely that iron (or

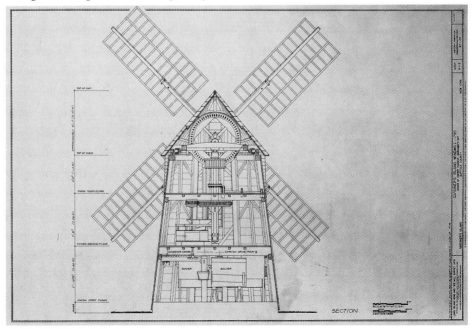

any another product) would be beneficial to him or society only if it were related to a successful, sell-through industry. And, despite the collapse of the Hammersmith Works, Winthrop's dream would eventually be fulfilled: by the time of the American Revolution, the American colonies, thanks in part to technologists trained at Saugus, were turning out more and better iron (30,000 tons of it annually), through better industrial techniques, than was the mother country. In fact, Americans were then producing more than a seventh of the entire world's output. Despite this early birth of the U.S. iron industry, and despite Winthrop's foresighted accomplishments at Saugus—which site is marked today by a National Historic Park reconstruction—this inventor/industrialist and his works are properly viewed as a premature phenomenon, a great and important exception to the primitive character of American society in his time.

In the New England colonies and in England's other Atlantic-coast colonies, precious few notable inventions could be recognized, adding up to a discouraging lack of improvements or industrial advancements on land and sea. Oh, yes, enthusiasts of early and native American ingenuity speak of the innovative clapboards for the houses of Cape Cod, the readily built split-rail fences of Virginia, the Conestoga wagons and Durham boats of Pennsylvania. But these, too, were exceptions to the static mood and primitive technology that generally prevailed.

"Undeveloped" and "requiring foreign assistance" were accurate words to describe America for nearly 200 years after 1607. Jamestown's first sawmill had been built not by an ingenious new American but by an experienced German craftsman; similarly, the first sawmill in the Massachusetts Bay Colony, at South Berwick (now Maine), was built in the 1630s by a team of helpful Danes. Then, as might be suspected, John Winthrop investigated the prevailing system and improved upon it; he built Massachusetts's second sawmill near Braintree in 1644. Curiously, these implanted and home-grown American sawmills—equipped with up-and-down saws in frames activated by waterpower—were being set up here at a time when England had no such industrial lumber mills.

An impressive, trans-Atlantic industry developed around New England's technological advances in woodprocessing, with carefully shaped masts for the British navy being the most desired product. As early as the mid-1670s, ten "mast ships" were departing annually from New Hampshire alone. Maritime historian Robert Albion reports that the business was so brisk that occasionally eager shippers in New England sent sail-powered rafts made of masts across the ocean to England.

What economic advantages New England owned as a result of the region's timber (and its fish, as will be recounted below), Pennsylvania matched as a result of its high-quality iron. Prised from mines not far up the Schuylkill River from Philadelphia, this iron was forged and manipulated into hardware for William Penn's colony. Though technological advance was slow in the mid-Atlantic states, held back as in New England by local suspicion of untraditional ways, a number of notable, wood-and-iron-age innovations developed here, mostly at the adept hands of German settlers. Vernacular inventions from Pennsylvania included everything from the balanced hammers and the Conestoga wagons mentioned above to "two-sided shoes," one of the clever ideas that blossomed in cosmopolitan, fashion-minded Philadelphia. It was into that latter, creative society that penniless seventeen-year-old Ben Franklin landed in 1723, having fled from indentures in Boston and having found nothing of interest in New York.

The pinion, or small driving gear, of the Gardiner's Island windmill is seen at the bottom of the detail photograph above. It was called the "wallower" by early millers; the view shows it meshing with the wooden teeth of the brake wheel. The detail below shows how shafts and wheels, designed and produced in wood, were reinforced with iron bands and rims.

The vivid 18th-century woodcut above depicts the rushing waters harnessed by an anonymous engineeer for the sawmill of Henry Livingston, Jr. near Poughkeepsie, NY.

Benjamin Franklin, shown above working at his brother's printing press, fled servitude in Boston to become, inter alia, an experimenter with a new source of power—electricity.

The very name Franklin bespeaks inventiveness: science mixed with practicality—plus a certain sense of humor. The best-known story about Franklin as a great mind willing to ponder all new creations occurred when he was in France immediately after the American Revolution. Floating overhead on that day at Versailles was the Montgolfier brothers' brand new invention, a passenger-carrying balloon. When one of Franklin's companions mocked this remarkable contraption, saying it was of little use, Ben retorted, "Of what use is a newborn babe?" His point about new inventions was made, for all to ponder.

Franklin's unique genius, as an American inventor, was that as well as combining the scientist's curiosity with the wood-and-iron tinkerer's skill, he focused on certain social objectives—lightning rods to protect citizens from fire, bifocal spectacles to help oldsters stay focused. Although many of Franklin's inventions and writings brought him glory, not all of his ideas were immediately accepted, a fact which failed to deter him from his purpose. Take the soybean, for example: after Franklin introduced it to this country from the *Jardin des Plantes* in Paris, American farmers ignored it for more than a century.

Franklin's purpose was not to gain fame or wealth from patents but rather to help advance what he called "common living." One of his memorable statements on that subject (delivered in 1788 when he, aged 82, had but two years more to live) ran as follows:

> I have long been impressed with ... the conveniences of common living, and the invention and acquisition of new and useful utensils and instruments; so that I have sometimes almost wished it had been my destiny to be born two or three centuries hence. For invention and improvement are prolific and beget more of their kind. The present progress is rapid. Many of great importance, now unthought of, will before that period be produced; and then I might not only enjoy their advantages, but have my curiosity gratified in knowing what they are to be.

Inventions do indeed beget more of their kind, as Jefferson observed. They tend to be a part of the society in which they occur. It's not surprising, therefore, that what turned out to be Benjamin Franklin's best known (and most highly promoted) invention—the so-called "Franklin stove"—was stimulated by association with the region's iron-makers. The German settlers, who had brought with them a legacy of cast-iron baking ovens, had demonstrated to their fellow colonists how much more efficient they could be than mere brick fireplace with warming "hives." On studying their handsome and cozy iron ovens, Franklin thought he could design a space-heating stove for the home along the same lines. This he did, using similar (but much larger) iron plates for the stove's sides and top; the unit was to be located within the fireplace, projecting slightly into the room, the smoke venting directly up the existing chimney.

The stove was a good beginning, and doubtless succeeded in keeping many families warmer than they would have been otherwise. But it smoked terribly, it needed to be improved repeatedly before it could meet its distributors' claims, and it was a modest contribution to the advancement of technology. Nonetheless, because of the success of later promoters in selling such perfected "Franklin" stoves and because of the magic of the name, the historic claim of Ben's stove—i.e., that it allowed Americans to advance for the first time beyond the mere burning of wood for the heating of their homes—is rarely challenged.

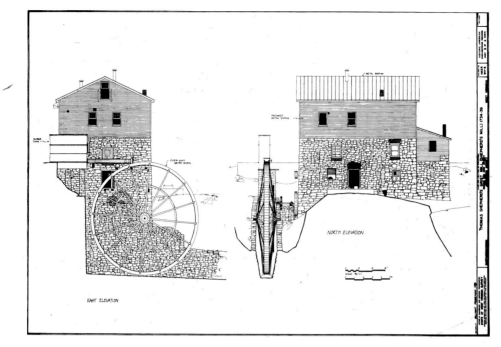

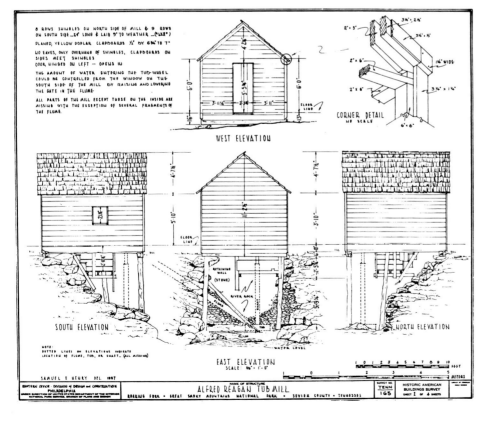

The early West Virginia and Tennessee mills at left show how rural Americans practiced alternative technologies. The upper mill was sited so as to make maximum use of water streaming across the top of the "over-shot" wheel; the lower mill was built to channel water beneath the building where it entered the "tub," or turbine, turning gears to grind grain on the floor above.

Nor were lumber and iron production the only industries that struggled into being in colonial times. In Philadelphia and a few other urban locations there were groups of capable, internationally recognized artisans who were cleverly fashioning brass and silverware from abandoned scraps and imported metals plus some native materials—an incipient arts-and-crafts industry, one might say. A number of particularly well-favored craft industries developed around Boston in the mid-1700s. But it must be noted that, generally, these cosmopolitan and creative craftsmen adhered to European patterns and production systems. More importantly, the whole industrial thrust of the next century was devoted to converting craft work into machine work, to destroying the crafts.

In traditional agricultural communities away from city centers, ingenuity in America's seventeenth and eighteen centuries was measured by such primitive devices as a water-powered grain-grinding apparatus. The invention, favored in the South, worked this way: A bucket suspended from one end of a long, wooden arm was filled and weighed down by piped-in river water; as a result of that action, the other end of the arm was raised, causing a mallet to rise up out of a bowl. Then the pivoted water bucket emptied itself, swung back up, and the weighted mallet crashed down to pound the grain. Ingenious? A little bit.

MECHANICS AND TINKERERS

It's small wonder that so many of these early colonial inventions tended to take place in the wood-working and farming industries, for that's where the money was. The first patent issued in America—1646 to Joseph Jenks in the Massachusetts Bay Colony—was for an improved sawmill and better-performing scythes. The possibility of mass-producing board nails by machinery, rather than individually by hand, continued to challenge the imagination and skills of ingenious Yankees before and after the

Even as late as 1857, when the newspaper illustration at right appeared, some wheels of industry—as at this ancient cider press—were still built in wood and turned by horses.

fall of the Saugus ironworks. By the end of the 1700s, some twenty-three patents had been issued for such machines. The best of them seems to have been the water-powered system devised by Jacob Perkins of Newburyport, a system which both cut the individual nails and gave them heads. So inventive was Perkins that, recognizing that his limitations lay not in his own ideas or brainpower but in capital to fund his creations, he ultimately emigrated to capital-rich England. Another highly creative Yankee was Jared Elliot, whose labor-saving seed drill caught the attention of change-minded colonial farmers in 1743.

Why all these creative Yankees? The remoteness of the upcountry New England villages made for a certain self-reliance, a need for farmer or miller to repair and improve things on his own. That situation, truthful of itself, may also be responsible for an ancient myth, disproved by many historians, the myth that Necessity is the Mother of Invention. Actually (as stated in Dynamic #2, above), the successful invention was usually the creation of a promoter, Yankee or otherwise, who knew what others had done and possessed a certain inventive streak which took him from there. With Elliot's seed drill and other, later agricultural inventions (particularly the McCormick Reaper, as will be demonstrated in Chapter Three) the myth was turned on its head: Invention was clearly the Mother of Necessity. That is, once the instrument existed, the inventor needed to sell it, and everyone needed to have one.

But Yankee tinkering did, irrefutably, lead to invention. Farm families had to make the best of their immediate surroundings, both for food and homespun clothing, mastering many skills, and developing that mechanical ingenuity that was to be so fortuitous for future industrial growth. Most artisans and laborers also combined farming with a trade, such as shoemaking, a trade which called for further tinkering. But most of this activity was for the community itself, with very little surplus for export or exchange with other towns and with little knowledge from other towns of what might have been a better way of grinding the grain or weaving the cloth.

The question was not how inventive was colonial America and how industrially forward but why (even with the exceptions named above) the backwardness existed and persisted among these obviously ingenious people. One answer, surely, was that the British Parliament imposed such restrictive pieces of legislation as the Molasses Act of 1733 (which was designed to kill the colonies' fledgling molasses industry) and the Iron Act of 1750 (which was designed to do the same thing to the iron industry), all in the name of keeping the empire's mercantile system flourishing. By the design of that concept, North America should supply only raw materials to the mother country. To invent, to manufacture, to sell abroad any up-graded item was illegal.

Another explanation for our lack of innovative productivity (an explanation that applied to New England, particularly) was that the earliest colonists had not come to these shores to create anything new. They had come here during the 1620s and 30s in flight from a tax-heavy economy and from tighter control by the state church, to preserve what they saw as the best of the old England. They thought like preindustrial Englishmen and they made their livings on farm or in city according to those beloved patterns. Nonetheless, despite the old-fashioned "peasant utopias" which historians now see as characterizing the New England landscape, the kind of mechanical genius described earlier did occasionally break through. Because each utopia was independent unto itself (as a matter of the Congregational Church's policies), creative individuals strove to find wider audiences. The outwardly directed Yankee mechanic and promoter—like Joseph Jenks and Jacob Perkins—gradually emerged.

Sociogeographers assert that the slowness of that emergence—all the slower in the South—may be ascribed to the seductions of our western lands, free lands made available in many areas as the Indians were forced off the known map by treaty or wiped out by disease. Those territories continually tempted Americans in the established eastern towns, making new enterprises unappealing. A cultural pattern developed in which, as young sons reached a certain point of discontent, they headed west. Rather than staying in the old town and coping with their economic problems (possibly by invention), America's ambitious youth took off for distant horizons, accompanied by rifles, Bibles, pregnant wives.

That wandering habit stood in contrast to the prevailing mode in Great Britain, where lack of land and natural resources forced creative men to stay where they were and seek greater power through mechanical devices. Great Britain, which had for the better part of the sixteenth and seventeenth centuries lagged far behind continental countries (particularly Italy) gradually turned itself around and became the leader of the Industrial Revolution in the eighteenth and nineteenth centuries. Invention, then powered by steam and coke rather than by wood and water, saved the day for the island nation, along with a superior understanding of the working of capitalism, as ultimately expressed in the writings of Adam Smith.

While Great Britain began to enjoy rapid industrial growth—thanks not only to the inventions of such geniuses as James Watt in steam and Richard Arkwright in textiles but also to the availability of investment capital (as a result of riches made in the wool industry)—eighteenth-century Americans continued in their backward economy, possessing no such capitalistic base. Their operations struggled to grow from the limited base of Native American agriculture (with some European improvements) and little else—with one crucially important exception: shipping. Mercantile fortunes made at the wharf-heads of the northeastern cities would enable Americans to get waterwheels turning, inventors dreaming, engines starting to do the work of man in more productive systems.

THE TECHNO-WONDERS OF MARITIME AMERICA

This beautiful whaleboat. Is it a work of art or an industrial tool? The answer: both. Furthermore, like other great inventions associated with major industries, the Nantucket whaleboat was a cause of the whaling industry's growth, not a result of it. That is, just as the balanced ax enabled frontiersmen to lead America's lumber industry into the first ranks, so the whaleboat created the whaling industry. Developed in the early 1700s, it changed what had been a rather haphazard, coastal occupation for English-speaking island settlers and their Indian colleagues into a manageable, seven-seas business.

Originally, whalemen had sailed forth from certain Atlantic coast ports and islands in heavy sloops some fifty or sixty feet in length, hoping to arrive in the midst of a pod of whales. They needed to be close enough to harpoon one of the beasts from shipboard—a rather infrequent occurrence. But the odds changed when the harpooner—often an Indian, as in Melville's Moby Dick—stood in the bow of one of the new,

WHALING IMPLEMENTS.

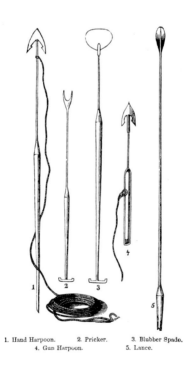

1. Hand Harpoon. 2. Pricker. 3. Blubber Spade.
4. Gun Harpoon. 5. Lance.

The Whale and His Captors, an 1850 volume in the Library of Congress, illustrates advances in whaling technology, from the early hand harpoon to the gun-launched version.

Illustrations in a book entitled Whale and Sea Fishery, *show a fragile whaleboat—one of that rich industry's greatest inventions—flung skyward by a South Seas behemoth.*

rapidly propelled, twenty-eight-foot assault crafts. These double-ended whaleboats sliced through the seas and rode the swells as adeptly as Viking ships. Light enough for a few men to carry, they were made of white cedar planks held in place by ribs that curved up from the keel. They just had room for the harpooner and four other oarsmen plus the mate. Specially equipped for its hunting purposes, the loaded boat was borne to the hunting grounds aboard the whaleship, lowered down into the waters when the whales were sighted ("Whale ho!"), and propelled toward one of the monstrous animals on orders of the mate.

Reaching the quarry, the harpooner would let fly. And then, after this first strike, the harpooner and the mate would switch places, athletically do-si-doing up and down the length of the six-foot-wide boat, the harpooner taking over the steering oar at the stern and the mate crouching in the bow as the wounded whale surged along at the end of the connecting line. Finally, the whale tiring somewhat, the mate would launch (in Herman Melville's words) "dart after dart into the flying fish," as the steersman sought to keep the fragile craft away from the "whale's horrible wallow." Melville went on to describe how "the red tide now poured from all sides of the monster like brooks down a hill. His tormented body rolled not in brine but in blood, which bubbled and seethed for furlongs behind in their wake."

Besides this whaleboat, this ingeniously effective, indigenously American craft, many other lesser inventions helped magnify the whaling industry—everything from the clever "toggle" at the head of the harpoon that secured the point in the hide of the whale to the "try-works" onboard the whaleships, the blubber processing vats which enabled the ship to be a self-contained factory-at-sea. But it must be admitted here again, as with the colonial iron industry at Saugus, that the whaling industry—which supplied a variety of necessary products, from the clear-burning flame of the spermaceti candles to the flexible baleen struts for ladies' dresses—only had so far to go. All the ingenuity in the world could not keep it advancing after it peaked in the middle of the nineteenth century, when petroleum lights began to be seen in most of the world's cities.

By then, however, the whaling industry had bestowed fortunes upon a number of families in Nantucket, New Bedford, Providence, and a range of southern New England cities. These were the fortunes that first started the wheels of industry going, mostly in the textile business. Even more impressive were the fortunes made in northern New England cities—including Boston and Salem, Newburyport and Portsmouth—from fishing and shipping. Here, too, maritime ingenuity played a key role, the evolution of the New England fishing schooner being the most spectacular development.

All through colonial times and into the nineteenth century, these schooners in both small, clumsy forms and in larger, more sea-worthy forms (but seldom more than fifty tons burden) risked the perils of the current-wracked, shallow banks beyond Cape Cod and Nova Scotia. There cod, halibut, and a host of other valuable fish were found—marketed, by the decade before the American Revolution, at the rate of more than $1 million a year. The business on the Banks was so valuable that this territory became known as New England's "silver mine." Ironically, the mode of selling the product—the far-sailing transport vessels—became even more important, in the long run, than the fisheries themselves. Increasingly large schooners, in whose holds fish were stacked and salted down for preservation, reached out to distant markets, transmogrifying coastal skippers into international traders. From this business of sailing abroad—to the southern colonies and to the West Indies as well as to Catholic fish-consuming Mediterranean countries—arose a new generation of merchants, keenly aware of their swiftly built vessels and of New England's economic potential.

In the years before 1700, Massachusetts Bay was by far North America's most powerful colony. Its population boomed instantly: By the end of the 1630's, Bay Puritans outnumbered the combined adult population of all other colonies in English-speaking North America, including proud Virginia. Furthermore, as that first colonial century matured, the merchants of Massachusetts shed any residual feelings that theirs was a settlement intended for religion. Clergyman John Wise, after surveying the colony's coastal trading ports reported: "I say it is the merchandise of any country, wisely and vigorously managed … this is the king of business for increasing the wealth, the civil strength, and the temporal glory of a people." That shift of emphasis from religion to trade, within a colony founded by Puritans, has long amused cultural commentators. Max Lerner, for one, sought to explain it by stating that "Protestantism and the spirit of Capitalism are virtually indistinguishable."

Another factor of special importance in that transition from religion to trade was the Crown's decision in the 1650s that the West Indies should be developed as a large-scaled, multi-island sugar plantation. Carefully constructed, newly configured ships would be needed to transport the product, ships built of the timber which New England possessed so richly. Thus Boston became not only the ship building capital but also the "mart city" of the West Indies, supplying fish and bringing back sugar and molasses for rum production. Thus, also, New England's canny merchants in their "appletree fleets" (so called because wood for the hulls was often improperly selected and cured) began to knit the Atlantic coastal colonies and the islands together in a not necessarily scrupulous shipping network.

Once embarked on that course, New England skippers found that the seas had no limits, that few laws needed to be respected. Indeed, the profitable business of shipping fish or wood products like shingles out to the West Indies or France and Spain, coupled with the shipping to England of goods from those lands and the bringing back from Europe to North America of goods desired in the colonies (including slaves from Africa)—this all settled into a free-wheeling pattern that has been dubbed "the Triangular Trade." Actually no triangle or any other geometric figure could ever be drawn to depict adequately the complexity, the inventiveness of this trans-Atlantic commercial trade. Altogether, it stands as a wondrous design, a brilliant chapter in the history of the age of mercantilism—which, we are instructed, led to the age of nationalism. Skippers became as adept at the international bargaining table as they were at the shipboard chart table; the flags under which they sailed became shifting signals of the development of respective markets.

The British reacted indignantly to the clever impudence of Yankee merchants. The sea lords knew well just how much smuggling, how much winking at restrictive, Parliamentary laws was needed for the whole thing to work. Nonetheless they recognized that, at the heart of the system lay the fast-sailing, far-sailing American merchant ship—a better and less expensive tool than they could bring into service. Surrendering to the superiority of these brigs and schooners, they themselves used increasing numbers of American vessels; by the time of the Revolution, at least a third of British merchant ships were American-made. More importantly, it was the growth of this fleet during colonial times and the experience of outsailing all other nations that gave this vast industry its foundations. Building on that, American ships after the Revolution would go on to Sumatra, to China and India, demonstrating the round-the-world capabilities of the young Republic.

REVOLUTIONARY INVENTIONS
FOR THE SAKE OF FREEDOM

This handsome, old hunting rifle. It really should retain its original name, the "Pennsylvania rifle," even though on the expanding frontier of the 1760s it was carried by men who called it the "Kentucky rifle." Reason for its primary name: Pennsylvania gunsmiths originally from Germany and Switzerland had perfected it, making so many improvements on old-world models that another, truly American invention came into being. As Jefferson would say, one idea had led to another; a new creation had resulted.

Before describing those improvements, however, it's well simply to look at this long lean, superbly wrought sculpture of mellow, polished, curly maple and gleaming brass hardware. Again, as with the balanced hammer and the whaleboat, one gets from this creation a sense of what the Greeks were talking about, beauty perfectly twinned with utility. Among the improvements worked out by Pennsylvania gunsmiths were the following: They had lengthened the barrel, thereby increasing accuracy; they had decreased the rifle's bore (believing that a smaller shot could fly farther); they had added an essential, brass patch box, to hold at the ready the greased pieces of cloth in which the bullet for the new load was wrapped. Now the weapon was ready for hunting whatever game might be spied, including, ultimately, redcoats.

Daniel Boone had spread stories about the amazing accuracy of this uniquely American weapon. Armed with it, a fair marksman could hit a squirrel 200 yards away, three times the accuracy of the average British army musket. Unfortunately, the rifle required a number of long seconds for reloading. Thus, as Boone told the tale, frontiersmen made sure they felled their game with one clean shot. The possibility of confronting an angry, wounded bear made for better marksmen.

The Pennsylvania rifle was deemed, in its time, the perfect weapon. British officers who encountered it slung across the shoulders of their colonial allies in the French and Indian Wars concluded that this must replace the musket in the royal army's arsenals. But their suggestions were refused, not merely because the weapon had no bayonet and took so long to load (the riflings resisted as a soldier ramrodded the specially wrapped bullets down into the barrel) but also because army regulations ordained that soldiers lay down a "field of fire" in front of them. That blast-everything procedure seemed far more important than pinpoint accuracy in the days when opposed armies addressed each other in close-order lines of battle.

When the American Revolution broke out, it soon became apparent that this Pennsylvania rifle, this most treasured of family possessions, was not powerful enough by itself to win the war. Indeed, our Continentals were generally forced to play the war game according to the small-arms procedures of the British: close-order lines of battle blazing away at each other with muskets from a distance no greater than a football field. Hence the bloody fields of Bunker Hill and Monmouth. But in many another engagement, mostly in the South, (including most notably the King's Mountain victory of 1780), the sharp-shooting rifles of recruited frontiersmen won the day with their accuracy and punch.

Though by no means the magic weapon hailed in retrospective tales of the American Revolution, the Pennsylvania rifle was an advantageous tool for the victorious underdog. Furthermore, this rifle, which came into prominence when military leaders were establishing industrial standards for fire arms manufacturing, served as a challenge for the future: could such a superior weapon ever be produced and replicated in great numbers?

By contrast, the American submarine looked like the secret weapon that could, of itself, win the war. Although small arms and artillery might do something to slow down the British assaults on land (and although the very distances of the land might exhaust the marching redcoats), no weapon or combination of weapons seemed to be at hand to hold off the crushing attacks of the British navy. None at all, until a young Yale graduate named David Bushnell came along with his idea of a submarine. After his successful demonstration of this device to Connecticut's Committee of Safety, hopes arose that it was, indeed, the weapon everyone had been waiting for. Hearts leaped at the thought of one proud British warship after another going to the bottom of the sea, thanks to the work of an underwater craft—against which there was simply no defense.

Bushnell, though young, was already known for his experiments proving that gunpowder could be exploded underwater; Eli Whitney believed him to be an authentic "genius." There being no earlier known submarine precedents for him to copy (though Leonardo da Vinci and others had fantasized about it), Bushnell went ahead and solved each problem, invented each design detail. As for hull shape, he chose the egg, his theory being that this natural form could best withstand undersea pressures. *Turtle*, as his vessel was named (because the egg-shaped hull looked something like two turtles' shells glued together) was engineered to descend to the proper depth below the waves as a result of water admitted through a foot-operated valve. She would move forward thanks to a hand-turned propeller in the bow; she would stay on course thanks to a compass and a small, hand-operated rudder at the stern. But how to see the compass in the stygian darkness of the depths? As with all other questions, Bushnell had the answer: a phosphorus lamp.

As for the explosive device, Bushnell's submarine would carry that on its back. He planned to plant this "torpedo" on the targeted warship via a connection drilled into the hull by an auger positioned at the top of *Turtle*, near the hatch. So, with the approval and high expectations of American naval authorities, the first submarine ever to see battle sailed forth. At her helm was a brave seaman Bushnell had finally recruited, he himself being regarded as too frail for the physical demands of the assignment. The selected victim was the mighty warship H.M.S. *Eagle*, from whose mainmast flew the flag of Admiral Lord Richard Howe, admiral of the fleet which threatened New York with total destruction during that grim summer of 1776. If only *Turtle* could sink *Eagle* as she rode at anchor in the harbor, the British might be forced to cease their encircling campaign against the city, and General Washington's army could escape to fight another day.

Cranking his way underwater out to Governor's Island, the seaman succeeded in locating *Eagle* and in positioning *Turtle* beneath that enormous hull. He then attempted, repeatedly, to screw his auger into her bottom—the attached charge of gunpowder would be exploded by means of a Bushnell-invented time fuse after *Turtle* had withdrawn. But though he applied the auger with all his might, the submariner could not penetrate the *Eagle*'s copper sheathing, a rugged outer skin which had been added by

Despite the need for faster and better built ships for peace and war, invention supplied few improvements. In the engraving at left, Philadelphia workmen lend traditional skills— adzing, hoisting by pulley, and carrying by manpower—to the completion of a frigate in 1800.

19

British naval architects to prevent the action of seaworms, not submarines. Frustrated, the seaman released the explosive charge and moved away. When he figured he had traveled a safe distance, he set off the charge, making a very satisfactory explosion. The blast was not close enough to the warship to cause her any damage. Nonetheless, *Eagle*'s officers, alarmed at the near miss, decided to withdraw to an outer anchorage, far from the range of any such crazy Yankee devices.

The next year, the American navy launched the intrepid *Turtle* at Philadelphia against the British fleet anchored in the Delaware River. Again the torpedo failed to explode close enough to the targeted warship. David Bushnell seemed to have failed utterly with his highly vaunted submarine (as did Robert Fulton, at a later date, with a more complex vessel). Yet this lack of success seemed to cause Bushnell no personal disgrace: He was commissioned a captain-lieutenant of sappers by George Washington in 1779, advancing to captain by the end of the war. In 1783, he was given the august position of Chief of the Corps of Engineers at West Point.

Nonetheless, the reputation of flunked inventor haunted him wherever he went. He took an assumed name and found a job teaching school; at length he became headmaster. But then, his identity revealed, he moved south and tried being a medical doctor in Georgia. Until his death at 84, he was always tethered to the failures of the amusingly named *Turtle*—a scornful destiny that would await other American unsuccessful inventors. Although this nation has always hailed its successful inventors, it has, with the heavy hand of pragmatism, made life hell for those who tried but failed.

INVENTING AMERICA'S FIRST INDUSTRIAL COMMUNITY

This beautiful old community, Paterson, New Jersey, above the falls of the Passaic River, has been through many phases of industrialization, including strikes and fires and depressions. But at this city's heart, hard to see, lies evidence of a grand creation. Stones of the walls for a three-tiered sluiceway constitute that evidence, the raceway having been designed to power America's first industrial community, initiated in 1791. This was all Alexander Hamilton's idea, called his greatest invention. But these few stones (which may, in fact, represent later alterations at the site) constitute a meager show. Who would figure, today, what former grandeur they bespeak?

Hamilton invented a new technological system to advance America—a system which, we can now see, had its flaws. He believed, in effect, that he could jump-start the American economy into the new industrial age by designing and establishing a successful industrial community. The audacity of the plan reminds us of John Winthrop's unsuccessful Saugus, built 150 years earlier; it also bears a family resemblance to the successful industrial complex built at Paterson 150 years later, immediately after World War II. What Hamilton had in mind, like the Pennsylvania rifle, was a distinctively American invention, an excellent creation for its own day that would show the way into the future. It was to be unlike any industrial communities which then existed or were being planned in England. This was a visionary construction, full of promise, doomed to fail.

In the years immediately following the Revolution, one of the greatest threats to the young Republic was its indebtedness. This was not merely a matter of the $12 million owed to the French and Dutch creditors who had financed the war, it was also the dreadful, confusing matter of debts owed to the individual states, something like

This grand tumult of falling water at Paterson, NJ was the power that Alexander Hamilton sought to capture and employ in 1791 for America's first major industrial development.

$25 million. In the brilliant, ambitious mind of Alexander Hamilton, then Secretary of the Treasury, this latter debt should be assumed by the Federal Government. He intended that the interest (and possibly the principle) be paid to the creditors from government revenues—which were then mostly custom duties but might also come from manufacturing or from taxes—and he believed that this debt assumption was a healthy, central business for the government to be conducting ("an indispensable element of national sovereignty," as one recent historian phrased it).

Hamilton had to fight with all his political abilities to win this argument (an essentially strong-government argument) in the face of the Jeffersonians, who favored a far weaker federation, with the states managing their own finances like independent governments. But win Hamilton did, thanks, in large part, to his deal with the Jeffersonians whereby the U.S. capital would eventually be moved from New York City (and, briefly, Philadelphia) to the District of Columbia, amid the swamps of the Potomac. It has been argued that, in this maneuver, Hamilton created our strong, central government, the functioning reality authorized and enframed by the Constitution.

But the key to the future, Alexander Hamilton believed, was the inventive genius of the American people, our "peculiar aptitude for mechanic improvements," as he expressed it. This quality, properly applied, would make the young American republic prosper. In his famous *Report on Manufactures* (1791), written when he was Secretary of the Treasury, he expounded the principle that invention-driven manufacturing industries in the North, linked to agricultural suppliers in the South, would supply enough economic power to make the country independent and mighty. He also saw to it that Congress put in place subsidies and tariffs to nurture America's fragile, essential industries. These were: the whale and cod fisheries in New England; hemp and cotton in the South; iron and steel in the central Atlantic states.

Hamilton urged Congress "to induce the prosecution and introduction of rewards and premiums" which would favor inventors, viewing new inventions as fundamentally important to the building of a national manufacturing program. Even Thomas Jefferson, who had initially dreamed of a strictly agricultural nation of farmers and mechanics, and who had considered manufacturers and inventors ("artificers") as "the panders of vice," finally agreed with this part of Hamilton's program. Jefferson confessed that "Experience has taught us that manufactures are now as necessary to our independence as to our comfort." Going much further, Hamilton took every step to encourage the mutuality of interest between American entrepreneurs and the U.S. government. The tighter the government's handshake with capital, the better for the nation, he preached. And as a massive expression of that belief, he backed the foundation of a corporation that would create America's first manufacturing center—Paterson.

Scorned for his traitorous siding with England in 1776, Massachusetts-born Benjamin Thompson, better known as Count Rumford, demands recognition as a stellar early American inventor. Profiled above in a European uniform, he worked in Europe on the phenomenon of heating and deduced that heat was produced by motion of particles. In later times he was remembered by effective stoves like the "Rumford Roaster," at right.

His vehicle for making this happen was the Society for Useful Manufactures (S.U.M.)—an invitation to America's booming capitalists to get even richer with the full cooperation of the U.S. government. S.U.M. sought first of all to enroll the country's successful speculators and merchants as investors in its program, then went on to attract investors from abroad, coming up with a total of $1 million to fund this specially favored corporation. The corporation (a definitely for-profit entity in which the government played a central role), would, by Hamilton's earliest plan, build a colossal manufacturing complex somewhere in the mid-Atlantic states.

The size of S.U.M.'s corporation was as awesome as the concept was inventive. It would be larger than the total stock assets of all joint-stock manufacturing concerns in the United States at that time; uniquely, it would create an urban community of variegated factories, all devoted to producing goods and profits for the central organization. In Hamilton's eyes the timidity and small thinking of American business in the 1780s had crippled the country and prevented its leap into the new industrial age. The twin devils that he chose to conquer in one blow were lack of capital and lack of a trained work force equipped with superior tools. In the new community, a host of first-class workers would surely be attracted by the splendid accommodations. Inventions would spring forth like dandelions. Government-guaranteed incentives would ensure the flow of capital. Furthermore, this massive and successful corporation, as well as benefiting its backers, would demonstrate to American manufacturers conclusively how wise it was to do things on the largest possible scale. This was Hamilton's own invention for American success: Think no small plans!

As Secretary of the Treasury, Hamilton had ample occasion to observe the flaws of the past and to conclude that the economy must now be managed differently. The 1780s had truly been a time of peril and disaster. Shays's Rebellion in Massachusetts, in which the authority of federal courts had been usurped by farmers discontented with government tax policies, had been but one alarming sign that economic illnesses were bringing about the destruction of the nation. Hamilton was also painfully aware of the rash of business failures, many of them brought on by undercapitalization, old-fashioned tools, and poor management.

There was, for example, the Hartford, Connecticut woolen factory which had been established amid much celebration in 1788. This was the factory which supplied the thirty yards of broadcloth for suits worn by Washington and Adams for all to see at their inauguration in 1789. In but a few years, the enterprise closed its doors, unable (because of its inferior looms, low levels of production, and lack of marketing prowess) to create a volume of business healthy enough to counter distributors of British-imported cloth. Hamilton predicted that the larger-scale factories planned for Paterson would be funded and managed on a scale sufficient to avoid such failures. Also, the government would provide protection from foreign competitors. The Secretary promised all listeners that S.U.M.'s industrial village would be blessed with "a moral certainty of success."

After surveying available sites, Hamilton determined that the Passaic Falls location would be ideal for the complex of buildings he envisioned. Sited on New Jersey's fall line, this impressive cascade was the third highest in North America east of the Mississippi; the potential power input was massive. The community would be named Paterson, after New Jersey's cooperative governor, William Paterson, who obediently pushed the land-use bill through the state legislature. Pierre L'Enfant agreed to design the necessary buildings in the appropriate federal style, emblazoned with eagles. Ever gracious, the French-born architect (who had risen from private to major while fighting in the American Revolution and had gone on to redesign New York's city hall as the first seat of government for the United States) wrote a flowery reply in answer to S.U.M.'s invitation. He responded that, having recently been called to design "the capital scene of politics" on the Potomac, he would now create "the capital scene of manufactures on the Passaic."

Operations commenced there in the summer of 1791, the dawning of young America's industrial laboratory, planned to surpass England's Birmingham, to put the country back on its feet, and to challenge the world. The site seemed to have everything needed: excellent water power; decent transportation facilities, via the Passaic River and Newark Bay to New York harbor; a handsome, new cotton-spinning factory, equipped with the most modern machinery; a countryside rich with minerals, both copper and silver being mined in the area.

To supplement the unfailing supply of ingenious natives, there was to be (as Hamilton envisioned it) an ever-running stream of well-trained talent from abroad. With one hand, Hamilton offered cash rewards to immigrants who had snuck out of their lands with industrial plans hidden under their jackets. With the other hand he offered enticements to large-scale investors, promising that this favored corporation would receive an unbreakable monopoly, tax exemption for uncountable years, and perpetual control of the waterpower passing over Passaic Falls. As portrayed by Daniel Webster some years later, "Alexander Hamilton smote the rock of natural resources and abundant streams of revenue gushed forth." Well … at least he smote the rock.

Hamilton's enemies (specifically, Jefferson and Madison) watched the whole venture through narrowed eyes; they just couldn't wait to catch him in making too sweet a deal with this banker or that would-be industrialist. They had every reason to believe that such illicit arrangements might take place, for Hamilton was no more circumspect in this regard than any of post-Revolutionary America's opportunistic strivers. These were the go-for-broke land speculators, the gamble-everything merchant shippers, the new-rich auctioneers and hucksters, and the fortunately-married widows who took particular delight in the projects of Hamilton, their handsome and patriotic young leader—their pied piper.

Yet Hamilton was by no means universally popular, absent the Jeffersonians. President Adams, during his term, couldn't stand Hamilton, who manhandled the administration. Adams made free with remarks about this elegant poseur of a colonel, calling him "that bastard brat of a Scottish peddler." Adams concluded that Hamilton's ambitious and grandiose designs sprang from "a super abundance of secretions which he could not find enough whores to draw off." Although Hamilton's close relationship with Washington—so close that both regarded it as the next thing to a father-and-son relationship—gave the audacious Secretary a certain amount of cover, in this Paterson activity he was out there all on his own.

And Hamilton gave the Paterson project everything he could supply. When it became clear that there were not enough native engineers to design the machinery for the community's textile industry, which was the prime industry on which he concentrated (possibly a mistake, considering that the automated loom had not yet appeared on these shores), he dipped again into the European labor force and recruited increasing numbers of foreign experts. Sadly, none of these toolmakers were up to the challenge of creating a water-power system that could harness the mighty Passaic for purposes of the projected industries. When the promised "bounties and premiums" from Congress failed to come through, he devised other incentives. But these were far less attractive. When the financial panics of 1791 and 1792 depressed the stock market and dried up his sources of capital, he abandoned all other duties to hurl himself into the management of Paterson.

But by 1795, the corporation had sustained such heavy losses that the community had to be abandoned and the now embarrassing concept written off. In a few years, Paterson became a ghost town, a humiliating reminder that America had neither the inventiveness nor the money to make itself into something it was not yet—an industrial power. Alexander Hamilton, like Bushnell, took the bad medicine manfully, going on to become a more mature and cautious leader (particularly in the "quasi-war" with France in 1798), before falling victim to Aaron Burr's dueling pistol in 1804.

Hamilton had not been altogether wrong in his concept; he was merely ahead of his time and subjected to certain limitations that would not later apply. Those limitations included an inadequate labor force and an incomplete transportation system— the former being caused by a slackening of European emigration during those years, the latter by the lack of canals and railroads. Here at Paterson, the dynamics of American invention (particularly #1, fortunate timing) could not be effected. But perhaps the biggest lack of all here was in Hamilton's understanding of the role of government: To take such business risks was, far more properly, the function of private individualists whose potential misjudgments would not threaten the nation.

This collapse of the 1790s, this failure at Paterson, meant, of course, that all further efforts by the government to propel the nation into the age of the Industrial Revolution would have to be curtailed. Industrialization would have to wait until sufficiently well-heeled capitalists arrived on the scene, accompanied by even more adept inventors.

SAMUEL SLATER'S
MIRACLE ON THE BLACKSTONE RIVER

In observing the early failures of American industry, many leaders of business and government decided that not by development through inventions or manufacturing but only by advancement in trade could the nation find an augmented place in the world economy. More and more competitively, shipping magnates in the northeast sent whalers to the poles, clippers to the Orient, slavers to Africa. As a result of these worlds won, New England became even more powerful than before the Revolution, even though only one out of ten New Englanders was associated with the maritime industries.

Tight fistedly, the new and old millionaires of Salem and Newburyport and New Bedford husbanded their fortunes in wharf-side countinghouses. Whenever they were sly enough to hedge their bets through investment elsewhere, they would usually look to England or the European continent for stocks or bonds or an enterprise that was double-safe. Investment in such projects as Hamilton had advocated seemed an obviously bad idea. Invention interested them not at all, involving something unknown, therefore a bad risk. Old-fashioned ships, old-fashioned ways of making money (and trying by bribes and other means to control the wild, Jefferson-influenced politicians in Washington) appealed much more than anything new—even when one of those old ways was dealing in slaves.

This reprehensible business flourished with special virulence in Rhode Island, which had neither the fisheries nor the interior resources of Massachusetts. If stripped of moral considerations, slaves might be viewed as not too strange a product for U.S. offshore shipping. Indeed, the largest part of New England shipping then consisted of goods from elsewhere, mostly because this land of itself had nothing to export except (in the words of James Truslow Adams) "men, fish, and rocks."

So, more and better built Yankee ships transported spices from Sumatra, porcelains from China, wines and manufactured goods from Europe across the seven seas, making profitable exchanges whenever and wherever possible. Slaves, in many people's eyes, seemed just another part of this off-shore, out-of-conscience business. The fact that many of the Africans were separated from their families and sold to dealers throughout the southern states seemed of no more importance than that ninety percent of the rum from the West Indies was sold to New England distilleries. These were both items for shipping—potentially dangerous items, to be sure, but was that not the moral problem of the purchaser rather than of the shipper?

The Brown family of Providence, led by brothers John, Joseph, and Moses, profited largely from the slave trade. The family had its own flotilla of premier ships, which, with a few additions, had served as Rhode Island's state navy during the Revolution and, in the role of privateers, had brought in handsome profits. Because of that flotilla, the family's mansions dominated the city. Beyond shipping, investments were made in but few other industries, namely in distilleries and the fledgling iron industry. The family, deeply conservative, supported the militaristic Federalists against Jefferson and expected the state's politicians to fight for shipping interests (rather than manufacturing) in both the U.S. Congress and on the high seas. To secure their slave-delivery business, they made strong connections with aristocratic families of Georgia and South Carolina.

Moses, the youngest brother, began to see in the structure of the family fortunes a pattern not only dangerously antiquated but also appallingly inhumane. Becoming a Quaker, and thus a pacifist, he declared himself actively opposed to slavery, both the institution itself and the transporting of Africans. To make up for that negative economic position, he vowed that he would, somehow, find another, profitable course for the family businesses—perhaps manufacturing, perhaps cotton spinning. He hired mechanics and experimented with spinning equipment, his mills powered by the waters of the Blackstone River. His brothers disdained this quest, their attitude based on evidence before them (including the failures at Hartford and Paterson) that there was little reason to believe anyone in America could make a breakthrough into textiles. England had dominated the industry of spinning and weaving for so many years, guarding effective inventions as if they were military secrets, that it seemed clear the game was permanently closed.

In the view of most historians of the Industrial Revolution, including Brooke Hindle, the first director of the Smithsonian's National Museum of History and Technology, this mechanization of the textile industry lay at the heart of the Industrial Revolution.

Although certain other technological advancements were vital (specifically, the perfection of iron production, the development of the steam engine, and increased precision in machine work), the spinning of cloth by machine and the weaving of cloth by mechanical looms were essential processes for any nation wishing to advance into the new age. In England, that step ahead had been taken in the previous decade, thanks to the genius of Richard Arkwright. He had been knighted in 1786 for his invention and perfection of the "water frame," a factory-wide system for spinning cotton in great quantities.

Arkwright's system not only amplified and replaced the French-originated "spinning jennies" (machines having more than one spindle, operated by one person), it also stimulated drastic social changes. Whereas the jennies, like the ages-old spinning wheels, could be housed in the back room of a farmer's house or in a small industrial location within an existing village, the spinning factory changed the landscape. Removed from the old center of village life, constructed at the river's edge, the factory attracted around it the ramshackle lodgings of the workers and the grand houses of the managers. In later years, in New England, this assemblage of buildings came to be called the "factory village;" the church would no longer stand at the village center.

The community Hamilton had tried to create at Paterson was a more complex version of that basic pattern, involving a variety of industries. In England, industrial communities grew out of control, into grim and gloomy industrial cities, the most monstrous of which (meaning the most productive at the greatest social cost) was Birmingham. In these locations, as opposed to a few new cities reformers were planning in both Scotland and England, little effort was made to provide adequate housing or amenities for the workers; instead, all steps were taken to keep the workers in a kind of perpetual bondage. It was precisely this situation that made Jefferson such an opponent of large-scale industrialization, if it ever could be transplanted to North America.

Moses Brown, while fully aware of the potential evils of rapid, unplanned industrialization, saw himself as a social reformer, interested in schools as much as in factories. (Two still-existing institutions, Brown University and the Moses Brown School, would benefit from his attentions.) He believed that, just as there must be a way to implant the new textile processes in America, so must there be a way to control the social consequences. Allying himself with another Quaker named Almy, he intensified his researchings, began acquiring experimental machinery, and joined the efforts of other would-be-industrialists to recruit knowledgeable immigrants from the mills and factories of England.

One such immigrant, attracted to these shores by proffered bounties from America, was a talented supervisor of textile operations named Samuel Slater, born in Derbyshire in 1765. Having advanced rapidly in the mill of Jedidiah Strutt (a partner of Richard Arkwright), Slater took a gamble on the powers of his memory and set about pigeon-holing in his mind the designs of all the spindles and gears of Arkwright's "water frame." His problem then was how to slip out of England, the emigration of textile workers being prohibited by law. He ultimately disguised himself as a farmer, made it safely through the inspectors to a packet bound for New York, and arrived there storm-tossed in the spring of 1789.

New York at the end of the Republic's first decade was a bustling city, pulsing with mercantile activity. But, as viewed by Samuel Slater, the few textile operations there—

Although he transported English weaving technology to the U.S., Samuel Slater is hailed as an American inventor—for both his adaptations of those systems and his influential life here.

Illustrations on this and facing page, taken from an 1836 biography of Samuel Slater, show the enormous effects of his "invention" of the American textile mill on the workplaces of New Englanders. Rather than weaving or spinning in their own homes, the women at right stand long hours by machines driven by leather belts from overhead shafts.

including the ambitious New York Manufacturing Company—were primitive and under-financed, not sufficiently dynamic for him. Slater then recalled a chance encounter he'd had on shipboard with the captain. The captain had spoken of a certain Moses Brown in Rhode Island who desired to hire an English artisan familiar with the Arkwright loom. Slater sent a letter of inquiry off to that Mr. Brown in a place called Pawtucket near Providence. The reply came swiftly, promising a full share of profits from any spinning mill that might result from their relationship.

Thus it happened that Samuel Slater suddenly and happily found himself in the land of mechanics: New England. People here spoke his language, the language of whirring spindles and waterwheel-driven machinery. Slater instantly took over as the man in charge (even though the details of the contract ultimately signed gave him only a *half* share of the profits); he informed Moses Brown that most of the Quaker's previously acquired machinery had to be thrown out. Then, assisted by local mechanics, he drew upon the resources of his remarkable memory and constructed within what had been a clothier's shop above the falls of the Blackstone River something that looked very much like one of Arkwright's looms. But when the day arrived for putting his machinery in motion, it would not move, or at least it would not move as intended, to any good purpose.

The disappointment was all but overwhelming. As Moses Brown (whose family continued to supply the quantities of capital needed) fumed and sought to placate his brothers, Slater labored day after day to discover the hidden problem, the missing part—all to no effect. Tradition tells the tale that thereupon, exhausted, Slater fell into a dreamy sleep, a sleep filled with images of the driving shafts and spinning wheels of his mill. And in those dreams he saw the flaw: An essential, connecting band was lacking on one of the wheels. Rushing to the mill at first light, he corrected the deficiency and, in a few hours, his machinery was in full operation. Within no more than two years, Almy & Brown was selling quantities of their superior cotton thread in all major cities, producing enough of the admirable product to satisfy distributors who

By Slater's "Rhode Island System," whole families were engaged in textile manufacturing: men and boys can be seen at various tasks in the scenes above left. Spun thread—visible in cyclinders in the scene below—was a first step in the complex process of making cotton goods.

supplied it to home weavers up and down the east coast. For them all, Samuel Slater was the inspired genius who made the revolution happen.

That's how tradition tells the tale of America's first textile hero. Industrial historians, however, have come to believe that, Slater having done what he could to recollect the spinning machine, left it to two native craftsmen—Sylvanus Brown, a carpenter and frame maker, and Oziel Wilkinson, a mechanic—to translate his memories into three actual dimensions. This they had done with skill and, whenever some critical element was lacking or appeared out of place, they made up for the defect with patient

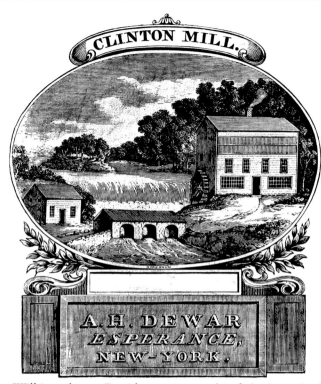

The Clinton Mill, as seen in this illustration for the trademark of New York's A. H. Dewar, represented the crude, if ingenious, structures found at isolated locations in the late 1700s.

imagination. Wilkinson's son David, carrying on his father's work, further demonstrated native ingenuity by designing, among many other inventions, America's first lathe with slide rest and lead screw. He also invented the highly efficient, evolved loom that would be used in most later American textile mills; for this accomplishment the grateful U.S. Congress granted him a $10,000 award.

The Wilkinsons, father and son, might therefore be credited with having created the ultimate machinery for America's textile industry. Thomas Jefferson would have said, perhaps, that they were the ones who, having absorbed other ideas, brought forth the "new creation." They gave the machinery the critical twists that made it work; they adapted it to the American scene. But the full dynamics of American invention were not with them: though they had the timing and the talent, they failed to possess the promotional dash and the capital connections. History tends to salute not the Wilkinsons but Samuel Slater (who, as it happened, married Oziel Wilkinson's daughter), often ranking him with Benjamin Franklin in the pantheon of America's great inventors. And he was, without doubt, an industrialist and social force of immense power in this period of America's industrial development.

Throughout southeastern New England Slater established spinning mills, instructing the new communities, as well, how to manage workers and worker housing. This was his so-called "Rhode Island System," an adaptation from English models; Moses Brown trusted that it would prove beneficial. At both the original Slater's Mill in Pawtucket (a reconstruction of which still stands) and at his other factories, the system called for the employment of entire families. Child workers' wages were only paid once a year, upon review of the youngsters' behavior. For old and young, the factory, the factory store, the factory-granted holidays would be the fixed focus of their lives. A kind of local peonage resulted, with everyone controlled by the ethos of the mill and its management.

Samuel Slater's intentions were charitable enough: For the children, he set up New England's first Sunday School, again following the pattern in England. But the whole thing, holy and secular, smelled rotten to some of Slater's fisher-farmer neighbors. Fearing the whole idea of an economy dominated by a single, obtrusive industry in the hands of a single rich man (however pious), these Rhode Islanders took it on themselves to destroy the dam Slater had built to hold water for the first mill wheel on the Blackstone—which dam happened to be, then, the biggest in America. After all, in conventional New England villages, had it not long been the practice for any citizen whose alterations of the land impacted on others to request the neighbors' approval before making such changes? That traditional, democratic step of polling public opinion Slater and the Browns had neglected to take.

But of course the dam on the Blackstone was soon rebuilt; the old economy was changing, men needed jobs, they could no longer afford such individualistic highjinks. And although many respectable citizens refused to sign up as laborers at Pawtucket and elsewhere—with only the poorest farmers and landless drifters consenting to lose their independence by becoming mill hands—this resistance, too, eventually gave way.

During the first third of the 19th century, the mill and factory—site of many an invention—had become the heart of a new community; here, the old center stands isolated in the background.

FROM WHARF TO WATERFALL

Following the lead of Slater's mill on the Blackstone River, factory after factory installed the American version of Arkwright's water frame. The making of thread for textiles advanced forever beyond the hand process, into the industrial age. Similarly, following the lead of wise and wealthy Moses Brown, other capitalists at water's edge shifted their investments from shipping to manufacturing. By this shift of New England capital, which resulted directly from the success of Slater's (or the Wilkinsons') invention, America stepped away from its agricultural base, stepped away from the sea, and stepped up to the next level—early industrialization in inland, riverside locations.

The dynamics of American invention had begun to assert themselves; Americans were choosing to improve their future chances by abandoning old lifeways and by picking up new tools and new systems in a way that was distinctly theirs. So you don't have a textile industry? Well, pirate the parts and create one. Make the new industry out of something borrowed, something old and trustworthy (like a mill wheel), something that perhaps has never been tried before. Experiment with electricity, like Franklin, for social purposes; get in touch with the thunder gods. Even before 1800, it had struck Europeans that this swift alteration of old world patterns, this ambition in the face of tradition, had become characteristically American. Wrote a Swiss visitor: "This is America, which dares to undertake the impossible."

And that spirit, which had been slowly evolving during the seventeenth and eighteenth centuries, and which suddenly boomed in the 1790s, helped give the nineteenth century nation its progressive personality. Ingenious Yankees, the British would call us. It seemed that, unlike any nation before us, we had deliberately left the past and our wood-and-iron origins behind us, if only to put all our energies into inventing an industrial tomorrow. As the historian Sidney E. Mead expressed it, "What Americans shared that bound them together was not the past but the present and a future."

*This 1810 drawing of an engine for the Navy
Yard in Washington, D.C., by one of Henry
Latrobe's draftsmen, shows the standards
demanded by that transplanted architect/engineer.*

TWO

INVENTING AND SELLING
THE AMERICAN SYSTEM

*C*onsidered this nation's first professional architect, Benjamin Henry Latrobe was born in England of American parents in 1766. Trained in Germany, he returned to England to establish a successful practice as both architect and engineer. Upon the death of his first wife in 1796, he moved to the United States—arriving in the middle of that highly creative decade. Having made a home for himself in Philadelphia, he contributed greatly to the architectural design and technological progress of the busiest city in the Young Republic, focusing particularly on steam-powered and hydraulic projects. He remained for many years a conspicuous example of America's need to be assisted by brainpower and professional competence from abroad, a galling reminder of our stalled development.

In a letter back to his brother in England, Latrobe wrote, "Here in America I am the only engineer.... What I do in my canals and buildings: I take Yankees, Jacks of all trades, and train them up to European standards." He recognized that at the very end of the eighteenth century Americans were showing some emergent native ability with tools; that recognition now seems condescending but was then accurate enough. Yet Latrobe had little idea of how, during the next century, Americans, as jacks of all trades and as inventors, would go through several metamorphoses and become national, then international heroes.

To some extent, Latrobe resembled Samuel Slater, in that he transferred European technology to America in such a way as to help ignite the Industrial Revolution on these shores. The hot fires of that revolution, the intellectual and material stuff on which it fed, were distinctly American—as were the results. As early as the mid-1820s, a British visitor here remarked excitedly: "The moment an American hears the word invention, he pricks up his ears."

Something very un-European was happening here. A large fraction of the population, particularly in the North, was leaving traditional ways behind, the people were eagerly anticipating new technological developments that would make their lives more productive, their homes easier to manage, their communities more amenable. Edward Everett, greatest orator of the era (and, later, the forgotten speaker who preceded

Abraham Lincoln on stage at Gettysburg), was heard to say of all these inventions: "What changes have not been wrought in the condition of society!"

To many historians of recent years, this materialistic turn (the time when, as Emerson expressed it, things got in the saddle and rode mankind) also marked the great and tragic differentiation between North and South. This was evidently the parting of the ways that would lead inevitably to the Civil War, with the agrarian South seeking to maintain its own "Cavalier" culture, as opposed to the industrialized and immigrant-packed North. Technological ingenuity served as the coiled spring that drove the northern states' new industrialism. Not just the minor works of small-town mechanics but large-scaled innovations by industrialists of vision transformed the northern landscape. Most important of those innovations was the breaking down of manufacturing into distinctive, repetitive steps and the logical arrangement of those activities on respective floors under one roof. This masterpiece of invention, repeated across the land, came to be called "the American Factory System."

Because of its wrenching social consequences, invention has sometimes been labeled as the northeastern neurosis which destroyed the nation's harmonious, agrarian order and split it in half. But invention was surely only one part of the whole nation-wide process of industrialization, which led to unimagined consequences. The steam engine, that epitome of the industrial revolution, found some of its most eager customers in Louisiana, where it went to work in the sugar mills. The process of invention, one might say, became a part of the early nineteenth-century American drive for individual and corporate wealth, a drive both creative and destructive. In the words of historian James McGregor Burns, the Economic Man had by then become the most important character all across America, replacing the Political Man of the 1700s.

Here in the United States the industrial revolution was also to take on a humanistic dimension that made it (and the social process of progressing by inventions great and small) something other than merely an advanced form of the upper class enslaving the lower. On the home front, the mass-produced clock on the mantel, the sewing machine in the bedroom, the newspaper with "telegraphed dispatches"—these American-made products from inventive American minds represented a unique acceleration of our nation's material culture. As well as exemplifying a middle class on the rise, they stood for a kind of liberation.

With progressive New York State (enriched by boomtimes after the miraculous-seeming completion of the 386-mile Erie Canal) leading the way into major constitutional reforms, nearly all white males all across America could vote by the 1840s. This beneficial connection between technological breakthrough and political advancement unfortunately had definite limitations, as well as frequent corruptions. Nonetheless, in the creative years when the Young Republic and the Era of Good Feelings and Jacksonian Democracy were opening up new vistas to native-born and immigrant populations alike (with the notable exceptions of the Afro- and Native American), the perceived intention of inventors and industrialists was, in the words of Franklin, to improve "common living." That creative intention remains one of the truly grand, if somewhat inconstant, themes of American civilization.

The role of the American inventor in society did indeed change in these years before the Civil War. No longer the anonymous tinkerer or the Renaissance gentleman of the age of wood and iron, he asserted himself as an individualistic striver, out for money and fame, as well as for certain presumed benefits for mankind. As the nineteenth century matured, he found himself working more and more in lockstep with large-scaled enterprises—an association which might also be regarded as exploitation. Though not allowed to be a free spirit (if that had ever been possible), he still yearned to have a voice in the shaping of the world around him.

THE DREADFUL PARADOX
OF ELI WHITNEY'S COTTON GIN

Eli Whitney, known to all as inventor of the cotton gin, stands as but one of those inventors who hoped to improve not only their own fortunes but also the human condition of their times. His results were mixed. Unlike many of the creative spirits encountered in this book, he had trouble with neither wife nor alcohol. But he did operate more as a late-bloomer than as an instant genius, and he was both a not-totally-truthful promoter and a disastrous manager of money.

Although a member of a distinguished Connecticut family, with a three-generations-long history in the state, Whitney could not summon the funds or the confidence to enter college until he was twenty-three. A higher education at that time (1788) was more customary for a would-be parson than for a technologist. In his youth Whitney had been an apprentice blacksmith and, like his father, a tinkerer with lathes and tools. Having made his own nail-making machine, he perfected the country's one and only hatpin-making machine. But he yearned to know more about the principles behind both technology and theology.

While at Yale, he volunteered to repair the college's valuable orrery—a delicate, English-designed model of the solar system. To the amazement of fellow collegians, he made the complex structure tick and spin the way it should. But, on graduation, what was his chosen profession? He couldn't hang out a shingle as "inventor," for no one would seek him out. He had entree nowhere; self-banishment to a possible teaching position on the western frontier or in the South seemed the only choice. He was now twenty-seven and, with his tall and heavy frame and gentle manner, already looked like an over-the-hill sage to many of his classmates.

In 1793, South Carolina beckoned to him with a teaching position at a hundred guineas a year. Whitney responded eagerly, sailing south on a coastal packet as soon as he could manage it. But again, in the South, he was odd-man-out: On arriving he learned that the salary was only half what had been stated. Angrily, he rejected the entire teaching profession.

Then he remembered that, on board their southbound ship, fellow-passenger Mrs. Nathaniel Greene (widow of the Revolutionary general) had invited him to come to her plantation near Savannah. Although he was morally affronted by the slavery system that underlay plantation life, he rode expectantly into Georgia with the intention of reading law in that state. He soon discovered that the most pressing problem (and greatest opportunity) in impoverished Georgia was not legal but agricultural: how to develop a marketable crop; more specifically, how to make a machine that could pluck lint out of locally grown green seed cotton. If that could be done mechanically, perhaps the languishing plantations would have a dependable product for sale. Dreaming of sharp-toothed machines for cotton cleaning, Whitney gave up all thoughts of law.

In this work, his objective was not only to aid the widow Greene (soon to marry a

Eli Whitney, when he posed in his later years for the painter T. K. Chappel, had finally won the wealth and security that had avoided him during his early inventive and promotional career.

Propaganda in the press: while smiling black workers grind out and gather cotton cleaned by one of Eli Whitney's first gins, plantation owners admire the marketable product.

plantation owner named Miller who had graduated from Yale only a few years ahead of Whitney) but also to salvage his own financial wreckage. Furthermore, he hoped by his invention to give the slaves, with whom he sympathized increasingly, some easement of drudgery. When all the cotton processing was done by their hands and fingers, ten hours were required for cleaning the lint from a mere three pounds of cotton seed. What might happen if human hands were no longer needed for that wearisome job? With the economic situation of the South so stagnant, and with slavery itself a questionable asset, one might even hope that the removal of a need for slaves to clean cotton would speed the end of that institution.

The "engine," or gin, which Whitney ultimately designed consisted of a hand-cranked drum which pulled the cotton from the seed through a sieve of wires; a faster-turning brush then cleaned the lint. Feminist historian Anne Macdonald has suggested that it must have been Mrs. Greene who gave Whitney the idea of cleaning the lint with a brush. Perhaps the gin was, like many another invention, the work of several minds. Whoever contributed what, Whitney's experimental machine did the job. In one hour the inventor cranked out enough cleaned cotton to match a full day's work of several workers.

So much praise and so many orders were then given Whitney that he concluded he and the Millers had a business on their hands. He would return to Connecticut to manufacture enough gins to meet the region's demands; perhaps that would help turn the South's economic situation around—for everyone's benefit, including that of the slaves. Local plantation owners, however, declined to respect the inventor's contribution, saying that many others had made more or less similar machines; they broke into his workshop to discover exactly how his worked. Planters also spread fast-growing

Whitney's gin was one of many devices used to process cotton. At left above, workers operate a water-powered engine; below, field hands manage a horse-powered cotton press.

green seed cotton over new and distant territories, since they would now be able to clean and market the harvest. For that process they would summon additional numbers of African workers.

Inadequately protected by patents and outmaneuvered by lawyers and representatives from the southern states, Whitney, by 1804, was deprived of any income and burdened with all the startup and production costs of the factory he had established.

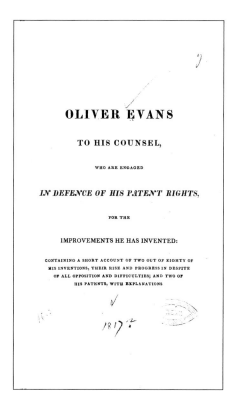

OLIVER EVANS

TO HIS COUNSEL,

WHO ARE ENGAGED

IN DEFENCE OF HIS PATENT RIGHTS,

FOR THE

IMPROVEMENTS HE HAS INVENTED:

CONTAINING A SHORT ACCOUNT OF TWO OUT OF EIGHTY OF
HIS INVENTIONS, THEIR RISE AND PROGRESS IN DESPITE
OF ALL OPPOSITION AND DIFFICULTIES; AND TWO OF
HIS PATENTS, WITH EXPLANATIONS

Having obtained one of America's first patents for his automated mill in 1790, Oliver Evans spent his life trying to sell the idea and then defending his patent rights and royalties.

Saved from economic disaster only by an act of Congress, Whitney at age thirty-nine was penniless and defeated on every front. His dream of bringing about the cessation of slavery had by then been totally dismissed in a paradoxical and dreadful way: The cotton crop was now earning the planters nearly $10 million a year; the value and the price of slaves, far from diminishing, had doubled; all thoughts of ending the system had ceased in the South.

Just as Eli Whitney had forsaken teaching and the law, he now forsook cotton production. Nonetheless, given the growing nation's continuing need for inventors, Whitney and his talents—and his ideals—would be heard from again.

OLIVER EVANS'S DREAM OF AN AUTOMATED FACTORY

Other struggling steps in the direction of labor-easing devices and systems for social good were being taken simultaneously by inventors in other parts of preindustrial America. Even as Whitney's gin was pirated and Hamilton's industrial village on the Passaic collapsed—in part because of the lack of a trained labor force—certain creative Americans, intrigued by the idea of machines more repetitive and precise than laborers, were attempting to automate (a word then unknown) both manufacturing and agricultural processes. Yet the idea of an amalgamated, continuous industrial process, what would become known as the "assembly line" in Henry Ford's time, seemed difficult to understand and accept at the beginning of the nineteenth century.

For example, there is the unhappy story of Oliver Evans, the Delaware farmer who created an automated flour mill in hopes of saving labor but who (having made a mill more efficient than any in Europe) reaped only scorn from his neighbors. Evans also invented and constructed a high-pressure steam engine that was far superior to the English, James Watt-designed, low-pressure engines that were then regarded as representing the highest standards. At first presentation of Evans's high-pressure steam engine, Benjamin Latrobe called the young mechanic's engineering theories "absurd." Ultimately, Evans was so discouraged by the nation's reception of his ideas that he entered a nearly suicidal depression and, as old age approached, burned all his papers.

The fifth of nine children, Oliver Evans had struggled against poverty all his life. At age fifteen he had been apprenticed to a wheelwright and wagon maker, studying mathematics and mechanics in his spare time. His master was so mean-spirited that he forbade the boy to use candles for nighttime studies. Evans had to collect wood shavings from his workplace; lighting those on the hearth, he scanned book pages by the flickering light.

The first indication of his inventive genius occurred in 1777 when, employed in the business of making stiff wire teeth for the carding of wool, he successfully designed a machine that turned them out by the hundreds each minute. Released from his apprenticeship, he joined his brothers in the building of conventional, labor-intensive flour mills amid Delaware's harvest-rich farmland. He soon found himself pondering the question of how such mills might be made more efficient.

It seemed to him that the bucket elevator might be at the heart of the process: He visualized many buckets arranged on a chain or a belt turned by gears off the mill wheel's shaft, carrying materials up or down, buckets rather than men. From that beginning, he devised ways to convey the grain and the meal between the several floors of

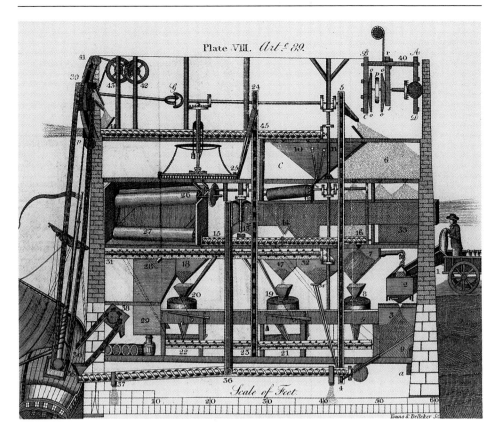

This plate from Oliver Evans's Young Mill-wright & Miller's Guide *shows the buckets, belts, hoppers, and screws needed to convey wheat from one end of the water-powered mill to the other.*

the mill at various speeds by means of other belts off the central power source. By 1785, he had structured all the newly invented parts of the process to work together. Whereas the usual complement of workers within a mill had been four men and a boy, now only one man (with an occasional helper) was required to run the whole system, from front door arrival of the farmer's grain to backdoor delivery of the ground meal.

One old Delaware miller was heard to say to young Evans: "Ah, Oliver! You cannot make water run up hill; you cannot make wooden millers." Sometime later—after Evans had obtained a state patent in 1787 for his fully integrated, automated mill—an inquisitive neighbor approached the mill on a snooping mission, only to find it whirring away, producing the product, with no person at all in evidence. Startled but not convinced, he and his peers concluded that Evans's mill was "nothing but a set of rattletraps, unworthy the notice of any man of sense."

The U.S. patent that Evans managed to obtain in 1790 was only the third issued by the newly opened U.S. Patent Office. Now he would surely earn a living from royalties. But they came in slowly, reluctantly, as millers in surrounding territories fought the logic and the legal obligations of his tradition-defying system. Eventually, particularly among the younger generation, Evans's position became more clearly recognized as his helpful articles guided fellow countrymen to a better understanding of the wonders of mechanics and automation.

A royalty payment received from George Washington at Mount Vernon must have cheered the inventor. In admiring French publications his mode of automation was referred to as *le système américain.* Nonetheless, members of his own family were free

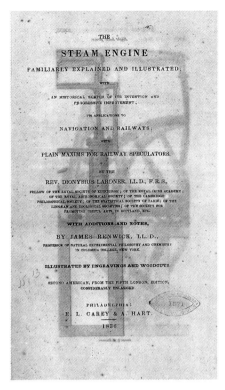

THE
STEAM ENGINE
FAMILIARLY EXPLAINED AND ILLUSTRATED,
WITH
AN HISTORICAL SKETCH OF ITS INVENTION AND
PROGRESSIVE IMPROVEMENT;
ITS APPLICATIONS TO
NAVIGATION AND RAILWAYS;
WITH
PLAIN MAXIMS FOR RAILWAY SPECULATORS.
BY THE
REV. DIONYSIUS LARDNER, LL.D., F.R.S.,
FELLOW OF THE ROYAL SOCIETY OF EDINBURGH; OF THE ROYAL IRISH ACADEMY;
OF THE ROYAL ASTRONOMICAL SOCIETY; OF THE CAMBRIDGE
PHILOSOPHICAL SOCIETY; OF THE STATISTICAL SOCIETY OF PARIS; OF THE
LINNÆAN AND ZOOLOGICAL SOCIETIES; OF THE SOCIETY FOR
PROMOTING USEFUL ARTS, IN SCOTLAND, ETC.

WITH ADDITIONS AND NOTES,
BY JAMES RENWICK, LL.D.,
PROFESSOR OF NATURAL EXPERIMENTAL PHILOSOPHY AND CHEMISTRY
IN COLUMBIA COLLEGE, NEW YORK.

ILLUSTRATED BY ENGRAVINGS AND WOODCUTS.

SECOND AMERICAN, FROM THE FIFTH LONDON, EDITION,
CONSIDERABLY ENLARGED.

PHILADELPHIA:
E. L. CAREY & A. HART.
1836

By 1836, when the learned history above appeared in London and Philadelphia, Oliver Evans's high-pressure steam engine had gained recognition. One of the working engines is pictured at right. But at first Evans's theories, opposed to the traditional British low-pressure engines, were considered outrageous.

with comments about how regrettable it was that Oliver's inventions kept him from "remunerative tasks." His father, early on, had called his son "cracked," and saw little reason to change the verdict as the young man's career developed.

Rather to the family's astonishment, Oliver Evans began to do well. He moved to Philadelphia, competitive Philadelphia, where he prospered as a merchant and distributor of mill supplies. In 1793 Thomas Jefferson (who had originated the Patent Office and had improved the Act in 1793, but who feared that patents gave inventors undemocratic advantages) took an active part in extending Evans's monopoly. To defend his patent position in the courts, Evans had to fight tooth and nail, winning legal benefits for all future inventors but bruising his own spirit. Having always been sensitive and aware of his relatively humble origins, he now became querulous, seeing enemies at every corner.

Benjamin Henry Latrobe headed the list. An adornment of Philadelphia society, he had designed the classically styled and well received new home for the Bank of Pennsylvania. He was also in charge of the city's remarkable waterworks, whose main pump was activated by a massive steam engine. This happened to be only the third

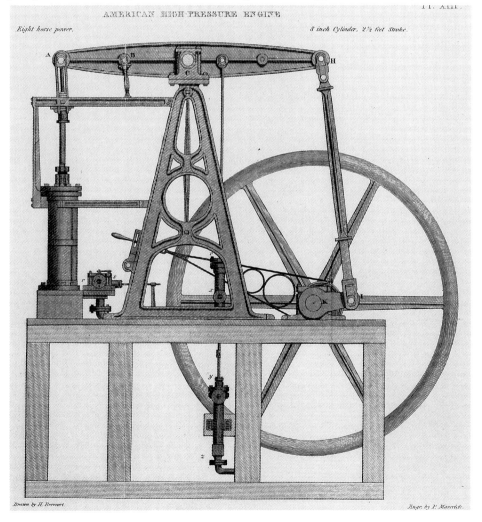

AMERICAN HIGH-PRESSURE ENGINE

Eight horse power.

8 inch Cylinder, 2½ feet Stroke.

Drawn by H. Brevoort.

Engr. by P. Maverick.

Unlike Evans's steam-powered, amphibious dredge of 1804, this slightly earlier invention relied on the energies of a horse and the good luck of a driver handling beams and cables.

steam engine operating on these shores in the 1790s; gigantic and dramatic, the engine had been specially imported from England. As with all British "atmospheric" engines, developed according to the classical theories of Newcomen and Savery, this paragon of an engine worked on the low-pressure principle, the piston being advanced by a vacuum in the condenser-cooled cylinder.

Oliver Evans, intrigued by the prospect of steam power, had wondered if more power could not be delivered if high-pressure steam pushed the cylinder. This way, without the need for a condenser, a more compact engine could be made; the engine could be at the same time smaller and stronger. The high-pressure steam engine which he then built could deliver a five horsepower thrust; Latrobe's low-pressure engine at the waterworks produced twelve horsepower, but its cylinder measured 25 times larger. Evans concluded that his was indeed far better and more useful—whether for transportation or for industry. By 1804, when he secured a U.S. patent, his engine had been improved to the point where it could deliver the same horsepower as Latrobe's. Would people now understand the importance of his inventions?

To make the point, Evans set about building a demonstration engine, sinking into it all the money at his command ($3,700). As described by Mitchell Wilson in *American Science and Invention,* the engine had a cylinder six inches in diameter, with a piston stroke of eighteen inches. The task assigned to the engine was to crush plaster of Paris by means of twelve jointly operating saws, an assignment which it completed handily—twelve tons of the stuff being ground up in twenty-four hours. Evans had made sure that his "great show" took place in full public view. But observers, while entertained, expressed doubts that the engine could do anything practical, such as grinding grain or propelling boats. Evans assured them that it could (and, in fact his high-pressure engines, a generation later, supplied the power for America's fleet of westward steamboats); nonetheless, the audience went away with their doubts intact. Who, after all, was Evans?

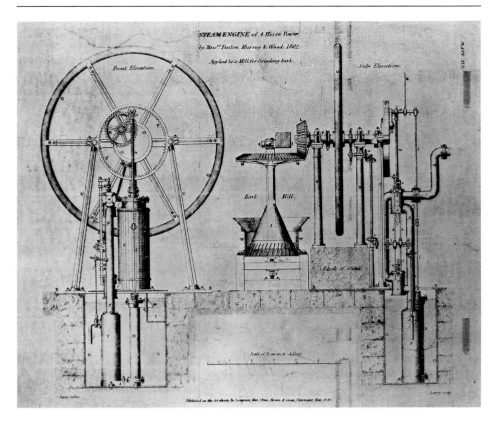

As the 19th century advanced, engineers and inventors found ways to use steam for many enterprises. The early engine above was applied to a bark grinder; its gears turned the "bark mill" at center. Most advanced of all was the heavy-duty engine below, designed by inventor George Henry Corliss at mid-century for New England's textile industries.

Once the city's most innovative possession, the steam-powered fire engine of Carlisle, PA seemed an ancient relic when this photo was taken three generations later in 1909.

Then, early in 1804, Oliver Evans made a rather spectacular appearance, center stage. In the strangest scow anyone had ever seen, he came steaming down the Schuylkill River to Philadelphia from his boatyard, propelled by a stern paddlewheel turned by a five-horsepower engine. Gradually, it became clear to astonished viewers that the whole, fifteen-ton vehicle was amphibian, for Evans proceeded to drive it ashore and direct it right into the city center. There he rumbled around and around Latrobe's waterworks in Center Square, soliciting contributions.

The vehicle was actually a steam-powered dredge, commissioned by the city for cleaning up the waterfront—Evans had named it *Oruktor Amphibolous* (the amphibious digger). Its engine hissed and thumped, the metal wheels chewed the pavement as mothers dragged their children out of the way. The inventor could not be ignored this time. But somehow, to his chagrin, the Philadelphians were unimpressed; they had, after all, seen John Fitch's steam boat on the river back in 1787—what was new about this?

What was new was Evans's far more efficient steam power plant, at which Philadelphians merely yawned. But in years soon to come, Oliver Evans scored victory upon victory. In his newly established machine shop, which he called the Mars Works, he and a crew of trained workers forged an increasingly impressive series of steam engines. These precisely milled, heavy-duty, "Columbia" engines for American factories ultimately helped transform the nation's industrial capacity. They helped the United States turn the corner away from the age of wood and iron into the new age of steel products and precise machines.

In his publications Evans kept trying to encourage young engineers across the new nation to follow his lead into this new age. If not in tradition-bound Philadelphia, then elsewhere, the superior quality of his engines became widely recognized. From

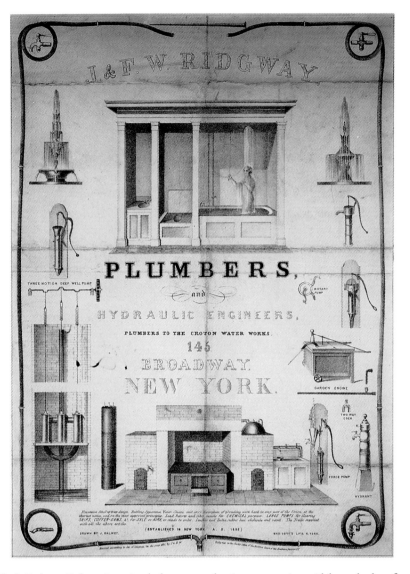

The prime feature of this 1835 ad by an inventive hydraulic engineer was the "water closet" pictured at top. It enclosed not only a wash basin and toilet but also a kind of shower.

New York Robert Fulton inquired about purchasing an engine. Although that famous showman's steamboat voyage up the Hudson River in 1807 was made thanks to the power of a Watt & Boulton engine (bought from England, with the additional payment of a fee that allowed its export), Fulton recognized the virtues of Evans's machines. On land, the first steam-powered industrial plant in the United States featured one of Evans's engines. This $15,000 piece of equipment, regarded as better than any power plant obtainable from England (given that it delivered twenty-four horsepower without fuss), was installed at a woolen mill in Middletown, Connecticut, in 1811.

Sadly, Evans's Mars Works caught fire in 1818 and disappeared in a pyre of flame and smoke. The inventor himself had but a year more to live. Having been called a "pompous blockhead" and worse for his efforts to get royalties due him (Latrobe's final comment was that he was "vain and dictatorial"), he became increasingly alienated and depressed. Yet, even as the courts helped him win settlements, he refused to go for blood, often declining to demand the triple damages he could have levied on recalcitrant millowners. Reminding his family of the "injustice and ingratitude of the public," he informed them that he had cast all of the drawings and papers for his inventions into the fire. He died an embittered veteran at age 64—still an outsider.

Invention by Clockwork—
The School of Eli Terry

By their innovations in two key industrial areas, textiles and steam power, Oliver Evans and Samuel Slater had made enormous contributions to America's technological advancement. But because of an odd twist in industrial history, the real breakthrough into mass production and factory-style simplification of the manufacturing process came not in those sectors nor in any other essential industries but in the utterly unessential realm of clocks.

Also, whereas inventor Oliver Evans and innovator Samuel Slater (along with industrialist Moses Brown and megathinker Alexander Hamilton) had certain social objectives in mind when establishing their mills and worker communities, such public concerns troubled the creators of the clock-making villages minimally. They operated in consonance with the family- and church-centered world that had always been. Typically, these millers-turned-manufacturers kept their operations as small and lowcost as possible; the mill itself was always designed (according to historian David F. Hawke) "in an image of sober rectitude." In these rather shabby circumstances the invention of assembly-line clock production took place (in its first phase), before the time-keeping industry had stepped out of the age of wood and iron into the era of higher quality, more precisely milled metals. Later, the industrial-age inventors who followed the early clock makers broke out of the village pattern, asserting themselves as individualists, determined to profit individually in more ambitious, factory-centered communities.

The initial break-through occurred in the same year as Robert Fulton's splendid event on the Hudson River—1807—the year when it appeared to many commentators that, in the perfected steamship, an inventor/promoter had finally given the nation the tool needed to move a populace across our continental distances. It was also the year when the U.S. frigate *Chesapeake* was forced to capitulate to the British warship *Leopard*, a scandalous international incident that helped bring on the War of 1812. The first step in the industrialization of clock making seemed far less dramatic: Early in 1807 the Porter brothers of Waterbury, Connecticut, approached a hard-driving, hard-selling clock maker in East Windsor and asked him if he could possibly make four thousand grandfather clocks in three years.

The clock maker's name was Eli Terry. Back in 1800 he had hooked up a waterwheel shaft to a series of belts in order to speed up the manufacture of clocks with wooden parts. This mechanization seemed to some (including the Porter brothers) a huge step forward, but Terry admitted that he had only "crude machinery" at his command. His ambition had been to make the sets of wooden parts for his clocks "by the hundreds," and he had more or less succeeded. But to make clocks by the thousands—that was something else. His neighbors thought he'd already gone crazy, to build such a dangerous, high-powered factory and to count on selling so many units.

Terry was a notable exception to the rule of conservative, tradition-bound Yankees. He was also very well tutored in clock making, having been trained by a craftsman named Daniel Burnap (who, as Benjamin Latrobe might have pointed out, had been

After Eli Terry's "inventions," every home and office seemed to need a clock. This wall clock, by Simon Willard, was the first ordered for the Supreme Court Chamber, in 1837.

trained by an English master). However well taught, they still seemed a poor lot, these upcountry clock makers, far removed in talent and accomplishment from the sophisticated craftsmen of Philadelphia and Boston. Terry himself might have been spotted on a typical day strapping three or four cheap grandfather's clocks to his horse's saddle, then riding up hill and down to find a farmer who could be convinced he really needed one. By contrast, the urban artisans, working in the patterns and traditions of Europe, usually importing their parts from the Old Country, produced precise, handsome masterpieces for the colonial and post-Revolutionary elite. How high a gentleman stood in the ranks of his port-city aristocracy could be told by the accuracy and elegance of the clock upon his mantel.

To the science of horology and to the creation of more precise clocks, Terry and his fellow New England clock makers contributed almost nothing. Theirs were generally untrustworthy time pieces, activated by easily warped wooden works. But, as perceived by men and women in the market which the clock makers addressed—a rural constituency in which, in 1800, only one home-owner out of fifty possessed a clock—such products were quite fine enough to occupy the hallowed place on the mantel. With Roman numerals handsomely set on scenic faces (always VIIII, never IX), the clocks delivered instant gentility if not accurate hours.

Again it should be stressed that these clocks as timepieces were not much of an invention. Nor were the wooden works a Connecticut creation (contrary to a well-established myth). New evidence shows that this technology probably came to New England by way of German immigrants in Pennsylvania. A certain Gideon Roberts, clock maker by trade, had lived for years in the part of Pennsylvania (Wyoming Valley) which was claimed by Connecticut in colonial times; when he moved back to New England, he took with him professional knowledge acquired from former neighbors who had owned good German and Dutch clocks, with wooden works. Soon after his return to Connecticut in 1761—particularly in the East Hartford area—clocks began to appear with works of oak, apple, laurel, and cherry wood.

It was not until Terry accepted the Porter brothers' proposal in 1807, and created a manufacturing system, that new things began to happen in Connecticut and American clock making. In business school terms, the Porters figured out how they could (by promotion and by immediate delivery of relatively inexpensive clocks to the country's new middle class) create a demand. A part of their marketing plan was to make the clock available for the bargain price of $20. And the salesman would stress that the first payment wouldn't be collected until his next trip out that way. How could you lose?

Although Terry's clocks were not that innovative, his manufacturing system certainly was. It seems to have been his own idea, highly unusual for that day. Having set up the necessary parts-making machines in a logical order, Terry assigned one villager to each work station; he or she would do only that task. No particular cleverness was necessary: Simply adjust the machine to cut out the same piece of wood the same way, again and again. The assembly and installation process was also simple and repetitive. Artists were necessary to paint the face and decorate the product, but generally clock making had become boring, no longer a challenging art.

So Eli Terry had the system and the capacity to handle the Porter brothers' order. And his system worked; Seth Thomas and Silas Hoadley (two clock makers who would later become famous in their own right) helped him instruct the workers and to complete the order on time. Both the Porters and Terry became rich as a result of the successful deal. Terry then saw how he could surpass that accomplishment: design and manufacture a clock that was even easier for salesmen and customers to transport; make it look even more elegant but still keep the price in the basement. Picking up the pace, he sold 3,000 clocks in the year 1809.

In 1814 he patented his 30-day, weight-driven "shelf clock", priced at $15, with which he would win a nationwide reputation, as inventor, as miracle maker. These clocks—many of them driven by the brass and steel works of the new machine age—were built by the hundreds of thousands in a new plant, "Terry's Mill" on Connecticut's Naugatuck River. There skilled mechanics improved the machines which made the more precise parts. This whole operation was a masterful industrial move; Terry made a fortune and created a surge of competition among would-be clock makers that has been likened to the Gold Rush. His well-labeled clocks, having been taken to all ends of this country, also found markets abroad (nearly annihilating English clock making in the process). The people who lost out in this machine-made triumph were the old-line urban craftsmen, the skilled artisans who had for so long kept clock making as their private preserve, new casualties of industrial-age invention.

In 1818 Terry improved his clock even more, again in 1822; in 1826 he won a new patent. He decorated the newest model in the "pillar and scroll style," popular for decades. To those who chose to share his glory with him, he granted permission to make their own shelf clocks in their own names, accepting royalties by way of thanks. He became "teacher to the trade" until his death at age eighty in 1852. Though many another clock maker went broke in the process of trying to create a similar empire, Eli Terry himself led a hard-working and successful life until the end of his days. As a kind of international tribute to Terry's phenomenal production and marketing genius, an English scientist, traveling though the United States in the 1840s, wrote home as follows: "Here in every dell in Arkansas and in every cabin where there is not a chair to sit in, there is sure to be a Connecticut clock." Forty years earlier, only the highest class of citizen had thought that he could afford such a marvelous product, ticking in time with the universe.

Entrepreneurs and leaders of other industries took note of Terry's contribution. What had been his big, golden idea, they wondered. Was it the mode of manufacturing or the nature of the product? And how strange that this breakthrough had occurred under those humble, upcountry circumstances, completely contradicting Alexander Hamilton's preachment that big merchant investment and a big industrial complex were the only way to crash into the industrial age! There seemed to be the possibility that ingenious, individualistic mechanics, with their eye on an expandable market, working precisely and delivering a novel product for a low price, represented the American future. Possibly American society would be liberated by means of their ingenious, indigenous works, rather than by Hamilton-style government programs or by Slater- and Latrobe-style imported technology.

The dream of riches from manufacturing had appeal throughout all American states, but particularly among New England capitalists. Squeezed by Jefferson's embargo against the British and by the War of 1812, these trade-rich merchants viewed further investments in the old, maritime enterprises with a sudden lack of enthusiasm. It was a tenet of the Yankee faith that "fortunes should not lie fallow." Something new was called for. Salem's tallest-standing plutocrat, Thomas Handasyd Perkins, for example, chose at this time to launch a new venture in China, transporting opium to that enormous marketplace of (as he saw it) heathens and pirates. But on the side—just in case—he also made a few cautious investments in various mills. For who knew? ... the rumors of new opportunities in those manufacturing towns might just be true.

As stated by historical geographer Jack Tager, neither industry nor agriculture would flourish on these American shores "until Massachusetts's maritime supremacy was put in jeopardy" by the foreign entanglements which had been brought on by policies of the Federalist and Jeffersonian governments. Only then would fortunes from the sea trades—fortunes even larger than those of Rhode Island's Brown brothers—be diverted toward the new manufacturing industries.

ELI WHITNEY AND THE MYTH OF INTERCHANGEABLE PARTS

Eli Whitney stands forth as another of those individualistic inventors who, in pursuit of his own fortune and destiny, found a way to change society. In Whitney's case, it all began somewhat as it had for Terry: He succeeded in obtaining a federal government order for a great quantity of muskets. The virtue of the Whitney muskets (as he stated in his proposal to the War Department) lay in the supposed interchangeability of their parts. The parts could be quickly manufactured, assembled, and repaired or replaced—or so Whitney claimed. He had made that boast back in 1798, at the frightening time for America when France, not England, had seemed to be the enemy. That was during the years when the no-longer-young inventor was attempting to recover from his cotton gin disaster. In the course of those legal battles, he had won many supporters. Among them stood Oliver Wolcott, Jr., a fellow Yale graduate and Alexander Hamilton's second-in-command in the Treasury.

To secure the much-desired contract for government muskets, Whitney had said that he could easily deliver four thousand fully assembled weapons in but a year and a half. Wolcott had backed him with all the Treasury Department's authority. The plan had seemed quite simple: he would try, just like Eli Terry, to reduce the work to a number of repetitive operations, the parts being machine made. Yet, in all honesty, Whitney had neither a building site for his factory nor tools to make the parts. And, within the range of his hometown of New Haven, there were scarcely any workmen trained to take on such an assignment. Nonetheless, he had sold the idea to the government on the basis of that one phrase, "interchangeable parts." It was a ringing promotional phrase—very important, ultimately, in the annals of American invention.

Yet today's historians, even at the Whitney Museum in New Haven, fail to back this American inventor as originator of the interchangeable parts concept. They suspect that it had immigrated to America from France, where French artillerists had viewed it as a desirable goal in military technology; also in France, the seductive phrase may well have reached the ears of Thomas Jefferson when he was serving as ambassador. Thus, when Eli Whitney used those buzz words again and again in selling his proposal at the capital, he was preaching in the highest places to those already converted.

Embarrassingly, Eli Whitney could not, did not, produce the four thousand muskets by his deadline of 1801. This he admitted. However, he still had to deliver some number of weapons as soon as possible. When commanded to come to Washington and demonstrate to President Jefferson and other officials that he had solved all manufacturing problems, he arrived and solemnly opened the cover of a grand chest. As all eyes watched, he picked and chose among the parts there gathered and swiftly assembled ten muskets. Voila!

The officials probably realized that this was simply a convincing way for Whitney to gain more time: The special weapons he had brought with him did indeed have interchangeable parts, but he had by no means perfected that system. His machines—the machines which were doing the work of men in his factory—were not yet that well calibrated. Not until 1808 did Whitney complete his order, sending to Washington muskets which were ultimately used in the War of 1812. They worked well enough, even if their parts were not completely interchangeable. [That development would not occur until a Maine man named John Hall, who had invented a breech-loading rifle in

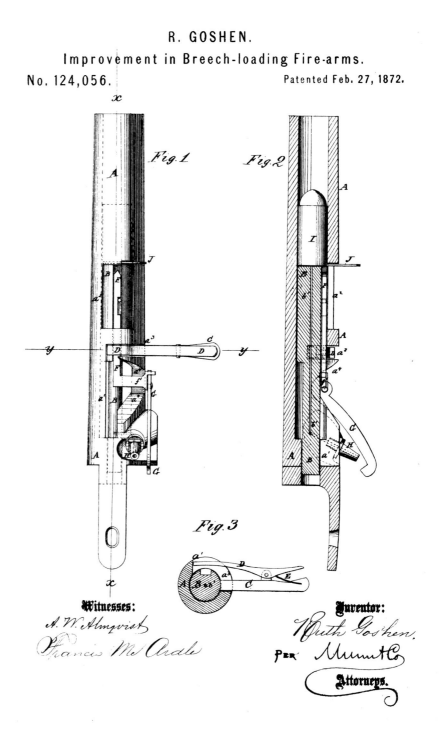

R. GOSHEN.
Improvement in Breech-loading Fire-arms.
No. 124,056.   Patented Feb. 27, 1872.

Fig. 1

Fig. 2

Fig. 3

Eli Whitney and other American inventors sought increasing precision in the milling of weapons. This patent for an improved breech-loading fire arm was issued in 1872.

Witnesses:
A. W. Almqvist
Francis McArdle

Inventor:
Ruth Goshen.
Per Munn&Co
Attorneys.

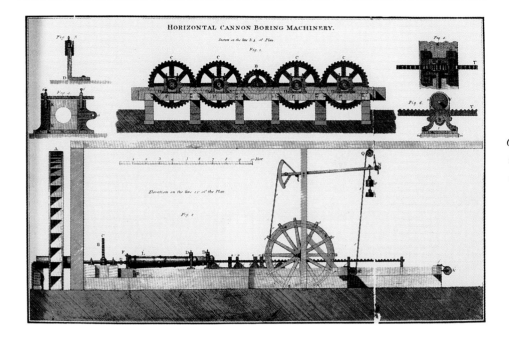

HORIZONTAL CANNON BORING MACHINERY.

Generous military budgets stimulated improvements in weapon manufacturing. The plan at left for cannon boring machinery is from The American Artillerist's Companion.

1781, established his superb milling operations at the U.S armory in Harper's Ferry, (West) Virginia.]

Soon after winning the unprecedented government contract, Eli Whitney had set about constructing his own factory town, now recognized as America's first modern industrial community. This he staked out at the family gristmill and farm at New Haven. Whitney was very aware of the ideals of the historic New England community, with its emphasis on modesty and godliness and cleanliness. But he also, as an individualist who presumed to reshape the world for his own advantage, approached the enlarged factory village with working practicalities at the head of his list. There would be, for example, no whiskey allowed on the job. This was no sloppy, barnyard operation; there would be, even, a vision of grandeur.

The new town came to be called Whitneyville, in the manner of the era soon after our Revolution when, despite international turmoils, every new worker-community had to have a *-ville* suffix to honor the enduring principles of the French Revolution. And because Whitney wanted to avoid the customary, shifting patterns of laborers moving independently among New England localities, he set out to build accommodations in which the workers would be permanently housed, as a part of their contract. By his system, female and male workers (as opposed to the Slater, or "Rhode Island," system in which children did much of the work) could then be trained with an expectation of their continuing residence and their sustained performance. With an eye to attracting a better class of laborers, Whitney himself designed attractive residences, to which the word "elegant" was applied by contemporary reporters.

A superior draftsman, Whitney focused his architectural talents and efficiency planning on the so-called "barn." Built in 1816, this handsome structure—with elliptical arches over the doors and windows—housed most of the factory operations and became the throbbing heart of the community. Whitneyville was also a place where major advances in milling techniques were made. Historian Sigvard Strandh credits

Whitney, at this location, with having created in 1819 America's first original milling machine, complete with multiple-edged cutting wheel and movable work bed (other historians credit arms-maker Simeon North with constructing the first milling machine, in 1816). Yet all commentators agree that here, in Whitneyville, where interchangeable parts and milling machines and assembly-line labor all came together in one new and coherent pattern, the so-called "American Factory System" was born. This was to be the highly efficient system which would allow the United States to surge ahead in the race for leadership of the industrial world. Though composed of many separately invented technologies, it was itself a triumphant and unique invention.

In 1809 a British merchant named David M. Dodge visited Whitney's factory and reported that it had

> a cheerful, tasteful appearance, like a tidy village. Every part [of each rifle being manufactured] … was made by machinery; any lock or stock would fit any barrel. Remarkable … no similar establishment in the [rest of the] country.

Mr. Dodge, having been sold on the uniqueness of Whitney's factory, went on to exclaim: "What a bound did my ideas make in mechanics [as a result of this visit]—from the operations of the penknife to this miracle of machinery!"

Another English visitor to Whitneyville remarked that this community, and those that would follow it, could be set up almost anywhere. His specific comment was that, by the American System, the factory could exist in an "extemporaneous town;" the mill and the worker residences (and indeed the community buildings themselves) needed for foundation neither church nor charter. Whereas New England mills had originally existed within the traditional, rural context (often reminiscent of the English landscape), something different was happening now in the generation between 1830 and 1850 . Find a likely stream on which that new mill—at first a cotton mill, then perhaps a harness-making factory—could be built, take the name of the biggest investor, add a "ville," and there you had, by the American System, the instant factory village. Strange new names began to pop up on state maps, names no longer reflective of past history or the motherland. In Connecticut, for example, there suddenly bloomed Thompsonville (carpet weaving), Collinsville (ax manufacturing), even Tariffville.

From Agricultural Inventions and Factory Towns to the New Industrial City

In one of its 1811 editions, the Windham (CT) *Herald* asked anxiously: "Are not the people running cotton mill mad?" And, to traditional eyes, there did seem reason for concern at the proliferation of new mills, new factory villages becoming towns. When Albert Gallatin, Secretary of the Treasury, toured New England two years earlier, he had counted some twenty-one competitive cotton mills. A few years later a correspondent noted that there were 170 of them in the Providence area alone. Surely, other newspapers argued, this was a healthy boom.

Agricultural boosters were simultaneously praising the progressive effect of the new country fairs, at which more scientific techniques were discussed and improved equipment demonstrated. They took pride in the force of invention: the new fertilizers and soil enrichers, the horse-drawn cultivators, the cast-iron plows (which had initially been rejected on the suspicion that the metal would impair fertility and "poison the

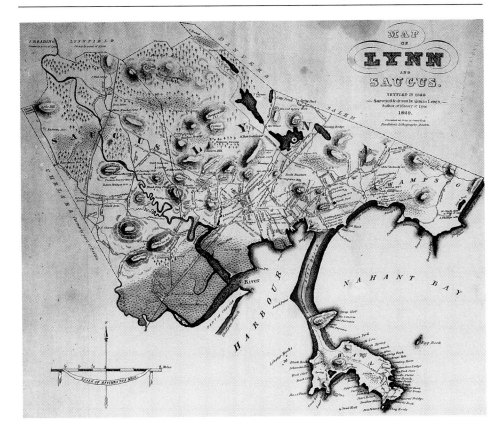

In 1829, before the invention of New England's industrial cities, Lynn, MA appeared in this gull's eye view as a pleasant harbor next to Saugus's iron works where U.S. industry began.

land"). Elkanah Watson, organizer of many of these instructive fairs in New England and New York State, was also energetic in advocating the new "spinning jennies" and looms for use in the farm wife's parlor.

Although the Windham editorialists might fret about progress and "mill madness," the majority of commentators found the prospect of productive little mill villages completely beguiling—and economically essential. Yale president Timothy Dwight (whose *Travels in New England and New York, 1821–1822* offers keen insights to the spirit of the day) had this to say about the scene: "To me there is something delightful in contemplating the diffusion of enterprise and industry over an immense forest." Harness makers, button producers, shoe manufacturers throughout the country sought to apply principles of the New England-invented American Factory System. As production increased thanks to those techniques, salesmen rode out to find new markets.

Near the end of this early period, in 1829, a Yankee writer named Zachariah Allen penned what remains perhaps the best description of typical communities. "The manufacturing operations of the United States," he wrote, "are carried on in little hamlets, which often appear to spring up in the bosom of the forest, around the waterfall which serves to turn the mill wheel." Zachariah Allen concluded that the United States would grow and prosper in this village pattern; American society would remain essentially rural and agricultural, with the family-managed factories providing energy and bright ideas at the center.

Yet it's worth noting that not even Zachariah Allen was totally sold on the idea of this agricultural-industrial Eden. He looked with some distrust at the increasingly paternalistic relationship between mill owner/store owner and the workers, noting that the store supplied millhands "with the necessities of life in order to retain them for service." He could see, perhaps, that although local customs remained intact within the centripetal New England town, certain new centrifugal forces were dangerously at work. As viewed by Marxist historians, mill hands now began to suffer the inevitable exploitation of one class by another. Since "social opposition was intrinsic to the nature of industrialization" (in the words of Jonathan Prude), the workers for many years to come would be linked in combat with their bosses.

Furthermore, while in colonial and early republican times agricultural communities had produced and consumed everything for and by themselves, now (as transportation became easier and shipping points flourished) outside markets began to determine what was manufactured or grown. That made for additional disruptions. Only certain products were called for, in gross and bewilderingly shifting quantities. Farmers had to respond to the whims of distant agencies, had to learn the head-scratching ways of capitalization and indebtedness; many farmers could not make the transition. As they drifted into industrial employment, these people of the land lost much of their independence. Textile worker Jabez Hollingsworth was quoted as concluding, "Manufacturing breeds lords and aristocrats, poor men and slaves." Hardly an atmosphere for free-thinking invention. Factory laborer Hiram Munger may have been quite accurate when he called such labor "slavery in the second degree."

Another jolt (perhaps well deserved) to the serene, static mill village was international economic rivalry. After the War of 1812, when lower-cost English goods again were allowed to flood the countryside, New England suffered an economic catastrophe. In the salty words of James Truslow Adams, "The collapse in prices and the sudden turn of the

Lynn fishermen had made shoes in their own homes since 1636. With the advent of shoe factories, workers lived in tenements crowding the waterfront, as seen in this 1895 photo.

Massachusetts textile mills planned not to employ children, in contrast to Slater's way in Rhode Island. That plan failed—as noted in this shot by labor photographer Lewis Hine.

consumer from domestic to foreign goods left the manufacturers as stranded and helpless as fishes caught above the tide line. ... Industry came to a standstill."

New England's cotton and woolen goods having been suddenly wiped off the market, America was once again almost totally dependent on British textiles. The shutdown of inland economies led to a heart-rending cry from New England for protective tariffs. It also led to the woeful national Panic of 1819, in which the commercial capital of Massachusetts dropped twenty-five percent, a heart-stopping depreciation. The stream of talented people flowing out of New England, leaving for more promising territories, doubled its pace and volume.

The rest of the country shed few tears at New England's plight. Had not the time come to put the Yankees in their place? For too long the imbalance had rankled: By the 1820s some sixty percent of U.S. power looms were in New England, with only three percent in all of Pennsylvania and Delaware. John Taylor of Virginia's Caroline County (who became known through his influential writings as "Taylor of Caroline") contentedly voiced the expectation that now the mechanics of New England along with their backers would be forced out of the controlling seats of power.

The financial down-plunge of 1819 so unnerved shaky northern politicians that, looking to the nation as a whole for aid and comfort, they distractedly agreed to the Compromise of 1820. In that inglorious document, authored by Henry Clay, the South won its way on the subject of slavery in Missouri (encouraging the corrupt hope that something could always, forever be done to let slavery and the cotton industry expand). Daniel Webster, called the Yankee Demosthenes, conspired with the Southerners in the interest of Yankee shipping magnates, seeking traditional ways to keep coastwise traders afloat and the republic from breaking apart.

After Welsh immigrant James Adams Dagyr invented a mass-production system of making shoes in 1760, each Lynn worker concentrated on one shoe part. This worker, confined to a work station beneath hanging fire buckets, attaches insoles to soles.

The next stages in industrialization of Lynn's shoe-making were the invention of wooden lasts for left and right shoes and Elias Howe's 1846 invention of the sewing machine (which replaced tacking or pegging). Here an inspector oversees a woman sewing parts of shoes together.

THE CREATIVE WORLD OF WALTHAM

While Eli Whitney stands out as inventor/creator of the first, small American manufacturing community, Francis Cabot Lowell must be recognized as the inventor/creator of the first large-scaled industrial locality—and thus progenitor of the new American city. Like Whitney, Lowell came from ancient Yankee stock; in his case, the line ran back to privateers who had sailed forth from Newburyport and other coves north of Boston to assault foreign ships before and during the Revolution. Nor did this bold young heir leave privateering behind him.

Francis Cabot Lowell, a trader in raw and manufactured cotton, had earned his first fortune by gambling the entire estate left to him by his father on eight far-voyaging merchant vessels. All came home safely and abundantly from their Asiatic voyages. He then went on, in 1808, to build Boston's huge India Wharf, a docking and warehouse complex which showed him to be something more than a fortunate princeling; he was a dynamic individualist, determined to reshape the existing world according to his precepts.

As he contemplated his next kingdom-building step, he looked far beyond the cotton-producing mill villages of southern New England and even beyond Whitneyville; they were obviously too limited and small-bore to free Americans from British textile domination or to propel U.S. manufacturing into a dominant world position. Yet the motivation behind Lowell's thinking was more personal than patriotic or geopolitical. He had a series of financial and philosophical objectives which grew from his very special Newburyport-Boston background; he yearned to steer his family and friends into a financially secure harbor where they could remain productive, prosperous, and slightly removed. His designs, though materialistic and profit oriented, reflected the moral, utopian tradition of his Puritan ancestors, plus a healthy lacing of piracy.

Taking two years off from his burgeoning operations in Massachusetts, Lowell voyaged to Great Britain, supposedly for health reasons. Actually, he hoped in all good health to learn certain secrets about England's large-scaled textile industry (which industry he, like Hamilton, viewed as the key to opening the way to future manufacturing kingdoms). He also sought to investigate a number of experimental towns in Scotland where something like social improvement was being attempted. Encouraged by the Scotland tour, Lowell traveled south to study, at the courtesy of his English hosts, the technology of their complex, formidable textile mills. Could that operation not be made more efficient by certain improvements, he asked himself.

Whereas, by British tradition, the carding and spinning and weaving of wool or cotton were carried out in separate locations, could not those operations be brought together under one roof, with immense savings? What Lowell had in mind was nothing less than a masterstroke of industrial espionage; he would replicate the machinery of British textile manufacturing at home, then he would reorganize it according to the American Factory System. Upon returning to Boston in 1813, the first concern of Lowell and his mechanics was to get the pirated, waterpowered loom up and running. This they attempted in a smallish brick building on the banks of the Charles River at Waltham.

At the same time, Lowell and his partners set up what would become one of the most powerful financial organizations in the United States, a group called innocently enough The Boston Associates. Given the continuing reluctance of New England's commerce-oriented banks to invest funds in manufacturing, Lowell invited close friends and family members to join their enterprise. In a way typical of Bostonians, whether they were overseas merchants or upcountry industrialists, this venture was going to be a "family affair." As he persuaded fellow members of Boston's upper class to invest in his manufacturing scheme (under the benign gaze of the Commonwealth of Massachusetts, whose encouragement of new industries made such corporations quasi-public), he gave his backers reasonable assurance that this exercise in "cautious capitalism" would yield a steady six percent return—without the old risks of adventures at sea. They could be quite comfortable "without the grinding anxieties" of maritime enterprises.

Investors leaped at the prospect. Lowell knew every one of them personally, about a dozen individuals; they endowed Boston Associates with more than $300,000. This hefty treasure chest was huge for that time (it was ten times larger than what the average New England cotton mill could command two decades later). Having analyzed the roller-coaster financial histories of previous, smaller-scaled milling operations in New England—many of which had failed for lack of long-term funds—Lowell decided that only one-quarter of this endowment should be spent for the mill, three-quarters retained as a safe operating margin.

Beyond that prudent financial program, there were other inventive strokes, both social and technological, in the total design for Waltham—the world's first factory to use power machinery to produce cloth from raw cotton within the walls of one building. First, Lowell and his associates integrated all operations (carding, spinning, weaving) on various levels of their three-story structure. The young industrialists also took a different, more inviting approach to labor (steady male workers in New England still being not that easy to find). They would recruit well brought-up women. "A fund of labor, well educated and virtuous" would thereby be created, brought in from the New England backlands. At Waltham these virtuous workers would be emplaced at their looms within a factory designed with large, light-giving windows—totally unlike the grim, English mills.

These youthful, single women, as well as being available and trainable, had the advantage to their employers of being neither tavern goers nor childbearers. Their wages would be in cash, unusual for that day, allowing them more freedom of choice in their purchases than had been possible under Slater's system of chits at the company store. Lowell expected that the women would keep their pay (set at a rate higher than that of contemporary schoolteachers), take the dollars home to their families, endow themselves with well-stocked hope chests for tomorrow's marriage. He seems to have had real hopes that, after a three or four year's employment, the "sisters" would go home as if returning from college, but richer. Like Eli Whitney, Francis Cabot Lowell wanted his factories to be as attractive, as healthy as possible.

The atmosphere at Waltham was notably less rural and more urban than in the mill villages—this was America's first industrial *city*. And from an economic point of view, the results of this experiment in labor and management were awesome: Waltham cloth sold so well that British imports and home-woven competition were nearly driven off the market. Domestically, the 1820s and early 1830s saw the end throughout the northern United States of the traditional domestic pattern, whereby most cloth had been made by women to whom the yarn had been "put out" for weaving at home on their own looms. After they had made the yarn, they would make the clothes. Looking back at her bygone farm days, textile worker Lucy Larcom remarked, "I somehow or other got the idea when I was a small child that the chief end of woman was to make clothes for mankind." Soon Lucy and her mother were buying curtains for a home that had never before known such frills.

The Industrial Revolution and its disruptions had arrived. In the case of the new industrial city, the revolution had happened not only because of clever planning and innovative management; it had also happened because of one, genius-filled inventor: The chief mechanic whom Lowell had recruited to help him start up the Waltham plant was a young mechanic, a "dirty fingernail mechanician" named Paul Moody. Born in 1779 and educated in a series of typical upcountry shops and factories near Newbury, Massachusetts, Moody had initially struggled to win acceptance within the tight fraternity of Massachusetts woolen weavers. He had then been recommended to Mr. Lowell.

There's the story of Moody's trip with Mr. Lowell down to Taunton, Massachusetts, a few years after the young man had been hired and had invented or adapted much of the mill machinery that would make the factories at Waltham hum. They had gone to Taunton to visit a seasoned inventor who had patented a superior yarn winding machine. Soon financial negotiations began, with Lowell being unable to persuade the man to come down from his unreasonably high fee for use of the new winders— even though Waltham would be using them on a large scale and the fee would therefore be multiple and enormous.

To make the point that the winding machines were essential to Lowell's operations, the senior inventor addressed the young mechanic who had come along. "You know you must have them, Mr. Moody," he said. But Moody had been studying the design with particular care. "I was just thinking," he mused, "that I could spin the tops direct upon the bobbin." The inventor, seeing that Moody was on his way to reaching an even better mechanical solution, spat out, "You be hanged!" Then he turned to the financier. "Well, Mr. Lowell, I accept your offer." Francis Cabot Lowell gave a little smile. "No," he said. "It is too late." And beckoning his ingenious assistant to come along, he headed out the door, back to Waltham. Moody's subsequent invention, known as the filling frame, was patented a few years later.

By that time Moody had been granted the title Chief Mechanic (1819), as well as what we would call today stock options; The Boston Associates sought to tie him as tightly as possible to their operations. Yet Moody, who invented for Waltham much original weaving equipment beyond the filling frame (most notably, dressing frames and roving frames), self-deprecatingly stated that his machines were but improvements on copied English originals. His contribution to the enterprise was, nonetheless, monumental.

As Paul Moody got the first mill at Waltham running and then engineered three other plants (with Francis Cabot Lowell helping him with the mathematical calculations behind the increasingly powerful machines), he worried that the flow of the Charles River was not adequate for expansion. Also, like the rest of New England, Waltham suffered economically from the War of 1812 and, after that war's end, from the flooding in of British textiles, leading to the Panic of 1819. But Francis Cabot Lowell had never been so rattled by the international threat as to become an extreme, tariff-demanding protectionist; on the contrary, he had argued for only the most "moderate" protective tariffs in the debates of 1816. The Boston company emerged unscarred from the economic crash (and even from Lowell's death in 1817), continuing to produce high-quality cotton sheeting.

Taking note of that steadiness at Waltham, the banks of Boston began to shed their prejudices. Perhaps manufacturing could be systematized, institutionalized, magnified; possibly investment in manufacturing would be as worthwhile as purchasing stocks in turnpikes or canal-building enterprises. The political and economic mood of all New England (not just of the grand fortune-holders) began to swing away from merchant shipping and toward large-scaled manufacturing, supported by improved internal transportation. By 1822—but eight years after the facility at Waltham opened—the more aggressive members of the board were calling for large-scaled expansion. They agreed with Moody that a far more powerful source of water was needed to turn the wheels of the extraordinary industrial complex that they now began to conceptualize. A grand city of workers and managers was projected, appropriate to this expansive era of good feelings.

Obviously the traditional mill village had been surpassed, it simply could not compete with Waltham or other urban production facilities. Yet editorialists expressed the hope (somewhat forlornly) that the advances in personnel management made at Waltham might now be carried on and even enhanced at the larger-scaled locations.

What no one could perceive was that, in some curious way, the gradual advances from the gristmill site to the mill village to the factory town and city, were already destroying the ethos that had, for centuries, held the community together in bonds of concurrence. The differences between rich and poor had become even more acute.

Furthermore, whereas the factory village had seen a certain number of regrettable accidents—millwrights crushed when attempting to install a larger wheel, fingers lost by undextrous weavers—now the community itself was subject to a high toll of more significant industrial accidents. An underclass was being created in the industrial city which would pay the penalty of factory-induced fires and explosions. Members of this underclass would benefit only peripherally from any technological advancements. The most notorious and revealing of these consequential disasters was the 1860 collapse of the mill at Pemberton, Massachusetts, in which eighty-eight workers were killed and 116 injured. Structural problems noted earlier at the mill had been deliberately ignored.

INVENTIONS IN THE SUPER CITY OF LOWELL

The route to the accident-prone, midcentury industrial city began when Paul Moody and a group of other mill-site explorers set out into the New England hinterlands in 1822. They knew of the broad and mighty Merrimack River, one of the region's two great waterways (the other being the Connecticut). The rushing cascade of the Merrimack's waters over thirty-foot Pawtucket Falls had been a boon to millers and a challenge to canal-builders ever since Independence. In the years since 1813—when the first sawmill/gristmill had grown into a textile factory for carding, spinning, and weaving fabrics—several little communities had grown up in the area. But Moody and his companions now reported that the river was capable of generating far greater output.

On reading that report, the late Francis Cabot Lowell's former partners set up the Merrimack Manufacturing Company, with Moody appointed superintendent. This company, with a capitalization of $600,000, would serve as financial vehicle for the founding of an immense industrial community. Where the ramshackle little town of

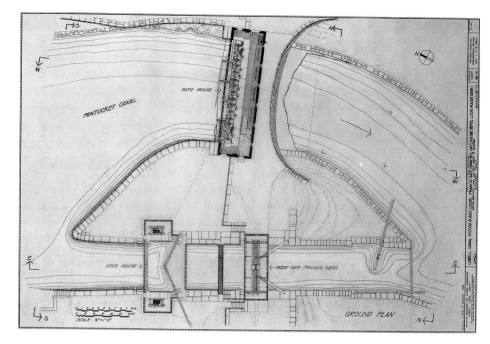

Engineer Paul Moody's first dam on the Merrimack at Lowell cost investors $50,000 in 1822. Diagram at right shows a set of guard locks on the resultant Pawtucket Canal.

East Chelmsford, Massachusetts, had previously struggled to survive, the carefully planned, industrial city of Lowell now grew. Within a few years it would become (at a working population of 20,000), the state's second-largest urban center. Operations at the plant began in 1823; by 1826, the Merrimack Company was producing two million yards of cotton cloth annually.

Paul Moody—for whom a family residence had been provided in the new city—again strove to find new ways to make machinery work for the greater profit of his employers. Given the force of water that would flow from behind the newly constructed dam, he invented and designed the world's first efficient, large-scaled water turbine. It powered the gears of the mill not by the horizontal drive of a vertical waterwheel but by the vertical drive of a horizontally positioned circle of encased vanes through which the water rushed. This far more efficient system, copied and improved upon by later engineers, became the standard power house of the typical, nineteenth-century textile mill.

Moody struggled to cope with two demanding, fulltime jobs under the orders of the "Proprietors" who had bought out the Merrimack Company: Those jobs involved both water power regulation and textile machinery improvement. In the latter capacity he invented a so-called "dead spindle," which was a revolutionary improvement on Samuel Slater's live spindle, imported from England. He also perfected the toplike, spinning "governor," a device to regulate the speed of the water-powered machinery.

Moody's most important improvement, however, was the introduction of lighter-weight shafting and additional belts to increase power generation. His weaving machines were thereby able to operate at higher speeds than their predecessors (the older machines having been dependent on slow-moving, heavy gears adapted from English models). In these operations, he relied increasingly on the skills of mechanics hired for and trained in his own machine shop (opened in 1824). Indeed, his machine-making and repairing facilities became so well known, that they soon became a profitable spin-off industry at Lowell.

62

Visible at Lowell today are ruins of locks, designed to make the best use of the canal's waters. That technology, originating in Europe's Renaissance, was improved upon by Moody and others.

But one small part of the total invention of Lowell, America's first planned industrial city, this vestigial set of worm and spur gears on Pawtucket Canal opened and closed a lock gate.

By this time, the dominant personality at Lowell was a certain Kirk Boot, precisely the type of hard-fisted manager needed by the new industrialization to separate it from the more human-scaled enterprises of yore. Boot bore the title Agent for the Proprietors and possessed the ruthless drive necessary to push Francis Cabot Lowell's concepts beyond their original bounds. In his thinking, the new city should offer power and advantageous sites to an ever-larger number of diverse textile mills; the whole would be served by a far-ranging system of sales agencies. The factories would be designed in the form of college-like quadrangles ranged along the water-supplying canals. Nothing would stop the city from being completed on schedule and on budget, neither the workers nor the flood-prone river. Boot expected his orders to be carried out as if they were military commands. With a variety of handsome brick buildings swiftly completed, and the deep canals between the structures reflecting bold towers and brilliant skies, the city soon won accolades as Venice of the North. Its output of textiles was phenomenal—by 1835 some twenty-two mills were producing 2.25 million yards of cloth each week. Lowell sheeting was then reaching even to China (which at mid-century, when the city had fifty-two mills, was taking 56 percent of American cotton exports).

In response to decrees and budgets issued by Kirk Boot and his management team, the mills at Lowell yielded far higher dividends for investors than had Waltham. This had the good result of encouraging expansion and additional investment (even as Moody discovered that the first dam held too small a body of water for many more mills). But it had the unfortunate result of encouraging the investors to believe that their six percent return was guaranteed, that it would never fail them even as competition grew within the textile industry after the first triumphs had been achieved.

A heavy price was paid for those achievements. When one of Paul Moody's most valued associates, Ezra Worthen, died in 1824, Kirk Boot added Worthen's responsi-

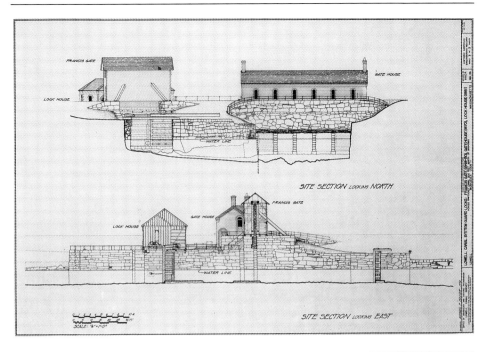

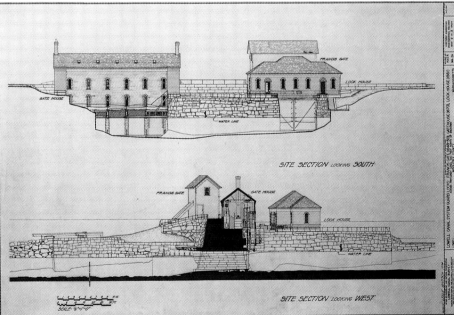

As Lowell expanded, its engineering grew more complex and daring. These two views (above: facing north and east; below: facing south and west) show the critical Francis Gatehouse.

bilities to those already on Moody's shoulders. During the next year, Moody upped the output of his central power drive to sixty horsepower, unheard of in that day. The entire effort was too much for the exhausted inventor. He died at age fifty-one after an illness of but three days in the summer of 1831, the victim (one might assume) of his own ambitions, but surely also of his employers' demands.

A stable, compliant labor force remained vital to the maintenance of Lowell's high production, healthy revenues, and steady return on investment. The young women of

New Hampshire and Massachusetts had, at first, seemed willing to respond to the call of the factory bell with unquenchable enthusiasm. Their letters home reported generally that the work was hard and boring but bearable. Literary societies blossomed; the long hours seemed acceptable because they made for good wages. But then, suddenly, those wages were cut (for, in a tightening economy, highest priority had to be given to the investors' dividends). In 1836, to the indignant surprise of mill management, the girls went on strike, a polite and reasonable strike, five years after Paul Moody's death from exhaustion.

This strike was by no means the first in New England among textile workers. The first had been a strike in one of the Slater-founded factories in 1824, followed by strikes in Griswold and Thompsonville, Connecticut, in the early 1830s. Compounding the fierce competition that had beset all factory towns was the nationwide depression of the 1830s. At Lowell the major issue was the fifteen-hour workday. The cry went up: Could the working hours not be reduced to ten?

The first Lowell strike lasted but briefly; the girls saw they had little to bargain with (even their parents urged them back to work). But these simultaneous actions, these unanticipated maladies which threatened the industrial city, indicated just how thoroughly the social balance of the mill village had now been destroyed—by the inventions which had brought so many benefits. There was truly no other arbiter now than the well-being of the gargantuan business itself; all decisions regarding workers or schedules or outputs were evaluated and determined only by their effect on the health of the industry. By now the pace of industrial and urban growth seemed to increase beyond anyone's reckoning: The working population of Lowell topped 32,000 by 1850, those hands tending 300,000 spindles and 9,000 looms. The mighty combination of capital and invention would expand the city even beyond those unprecedented numbers.

ENGINEERING THE NORTHERN INDUSTRIAL COLOSSUS

The complete exploitation of the Pawtucket Falls at Lowell on the Merrimack called for an engineering talent greater even than that of the mechanical genius Paul

No Niagara, the thirty-foot cascade of the Merrimack River at Pawtucket Falls was mighty and wide enough to provide power as Lowell grew into the world's premier industrial site.

Moody. Needed was an engineer who could not only design a large-scaled hydraulic system to control the power of the river but one who could also channel that power most efficiently into a variety of workplaces. Furthermore, this engineer, in obedience to the "lords of lash and loom" (as the textile bosses came to be called by the radical politician Charles Sumner), was asked to forget about utopias. These mills must let machines do the work of as many men and women as possible; workers must be viewed as subordinate and expendable.

As occasionally happens in human history, a man came forth who, while fully capable of answering that engineering challenge, had a mind of his own. This was by no means a Yankee individualist like Eli Terry or an exaggerating promoter like Eli Whitney. This was a thoroughgoing professional who, like Moody, would work within the confines of the corporation. Yet he would not allow his inventions to endanger the city which he empowered. His name was James B. Francis, undoubtedly one of America's most remarkable and responsible inventors and engineers.

He had been born in England, son of an engineer and railroad surveyor in Wales. Having learned some civil engineering at his father's knee, young Francis came to the United States in 1833, seeking wider opportunities. Fortunately he was able to land a job at Lowell, in the chief engineer's office. So assiduous was his work and so obvious his talent, that upon the death of his superior, Francis was put in charge of that office—when he was only twenty-two. It was to this surprisingly young man that the massive task of creating the enlarged industrial city of Lowell was given—an assignment which, when fully understood, required him to design a model for the rest of industrialized America.

Within but a few decades, more than fifty percent of Massachusetts and Rhode Island would be urbanized—a condition into which these states would lead the rest of the United States. By means of this rapidly spreading pattern of industrial urbanization, the northeast would for a moment regain leadership throughout the American continents. The political victory which had been scored for the South by Jefferson's victory back in 1800, when the axis of power had switched from Boston-Hartford-Philadelphia to New York-Richmond-Raleigh, now suffered a reversal. From now on, such southern leaders as John Calhoun of South Carolina realized that the economic power of King Cotton, great as it was, would not be sufficient to protect southern interests in the growing nation; they would have to develop an aggressive political program (particularly toward the lands of the West) lest they be crushed by the technologies of the new Colossus. At the same time northern senators began to pay less heed to port-city cotton shippers and to work more strenuously for the benefit of riverside manufacturers.

Francis's design for the enlarged, pace-setting city of Lowell called for a complex to be built on upper and lower levels, each supplied by water from different canal feeders. His objective was to guarantee a large enough water supply to the millsites so the manufacturers would never be subjected to the seasonal dry-ups that had cursed smaller locations. This he did, first, by rebuilding the original Pawtucket Dam (whose cost in 1822 had been a horrendous-sounding $50,000) which supplied the southern portion of the city; then he designed and built a wholly new dam and canal system, the Northern Canal (at the awesome cost of $551,584.70) to supply the new and northern portion of the city.

Photo at right, below, shows the view upriver from lower locks; note the factory stacks on the horizon. Proud Boot Tower (above) named for manager Kirk Boot, symbolizes Lowell's grandeur.

This Northern Canal posed a formidable engineering challenge, the greatest Francis would ever face. To tame the river and bring in sufficient water, the dam had to be more than three-quarters of a mile long (4,373 feet). The canal was 100 feet wide, with depth ranging from fifteen to twenty feet. The Great River Wall that rose between the Merrimack River and the canal was built of interlocking granite blocks laid without mortar; viewed today, it must still be judged an engineering masterpiece. Credit also belongs to the workers, for whom being drowned in the river or crushed by the rock was a constant dread. Most of the workers were newly arrived Irish immigrants, paid at the rate of eighty-four cents a day for a thirteen-hour day.

The whole new design was completed in 1847, with a great rushing in of waters and pounding of machinery. The increase in "millpowers" effected by the building of the dams (and by tapping the major lakes of nearby New Hampshire for additional water) was all that Francis and the managers had hoped for. More than enough flow was assured for the nine expansive enterprises which eventually grew along the never-failing waterways. And as Francis's mechanics improved the power delivery of the hydraulic machinery, less water was actually needed for each station. Key to those improvements were the "mixed flow turbines" which Francis invented to convert the waterpower to a drive system throughout the mills. Though turbines would continue to evolve, those that were then put into operation at Lowell (known by a variety of names, including Francis turbines) would long represent the world's most efficient systems.

To the shocked displeasure of management, James Francis frequently expressed concern about the safety of the total system, the likelihood of accidents. He was particularly worried about times of flood—which had been even more threatening to the small mills of New England than times of drought. Wasn't his wall massive enough to withstand floods, Francis was asked? Yes, but that wasn't quite the point, the engineer

By 1875, when the Pawtucket Dam was enlarged again (left), techniques for subterranean construction had not advanced far beyond those pioneered by engineer James Francis.

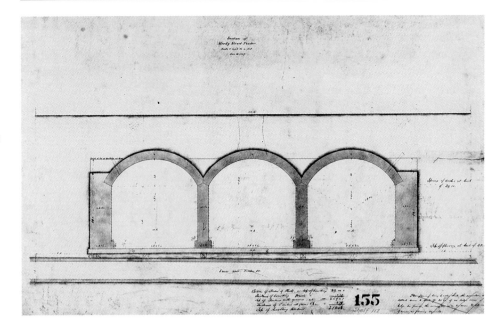

This plan for the Moody Street Feeder, built in 1847, can be found in the Historic American Engineering Records. It demonstrates the architectural stylishness of Lowell's creators.

replied; the flooding waters, directed into the heart of the city, might rise above their canal walls and accidentally overwhelm the populace.

Reluctantly the masters of Lowell consented to the expensive construction of a flood gate which they contemptuously called Francis's Folly. It was not only the cost of the structure that they resented; it was the idea that the corporation might have social responsibility. Had it not by now been determined that the people, for whom work and housing were provided (as well as libraries and other amenities), were subservient to the interests of the factory? In nearby Dover, New Hampshire, for example, workers had to sign a document accepting by contract *any* wages named by management. And, if that work involved occasional risks—well was that not the nature of modern industry? And was it not the role of the inventor to do simply what he'd been told?

Whatever bottomline sense there may have been in management's argument against his safety system, Francis felt fully exculpated for the expenditure by the events of April 23, 1852. As river waters surged fourteen feet above their planned level, his flood gate functioned as designed, drawing off the excess and saving the city. Taking his license even further, he invented and perfected a sprinkler system to assure against fire. Here was an American inventor who recognized that the new industrial city was, perhaps unintentionally, a mess of accidents about to happen. He did what he could to tame the combustible city which the corporation had ordered the inventor to create for the investor's benefit.

INVENTIONS FOR A NEW SOCIETY

But five years before the opening of Francis's enlarged canal system, Charles Dickens had visited Lowell—one of many Europeans who came to see this industrial colossus of the New World. Dickens, social reformer-novelist, generally approved of what he viewed in the new city, expressing only slight skepticism about this swiftly grown rival to Birmingham. His mistrust of the "Christian" intentions of the management and the

Inventions in other sectors teamed with textile advances to industrialize America. At left, workers form a tire—a result of the vulcanization of rubber by Charles Goodyear in 1839.

willing response of labor was, unfortunately, quite warranted: In the following decades wages were reduced repeatedly; strikes became bitter, endemic, occasionally successful. The estimable educator Horace Mann, who visited the city at about the same time, regretted that the industrial worker had become "a slave at morn, a slave at eve." Nonetheless, whereas less than a twentieth of Americans had been urbanized in 1820, the population of industrial cities represented a twelfth of the total U.S. population by 1840.

The question remained whether this was a change for the better—and what might be done if it were not. Within Boston's financial community there was also a split brought on by those who, having seen the supposedly benign communities of Waltham and Lowell become "haunts of misery and vice," now turned in other directions as more likely avenues for their investment, if not their social concerns. They considered railroads, oil wells, even rubber overshoes as perhaps more dollar-productive and less troublesome. Emerson and the Transcendentalists of New England also rued the consequences of pell-mell industrialization, taking to cabins in the woods for contemplation or escape.

Contributing to the unease in the Northeast were new generations of immigrants from Ireland and central and southern Europe. As "slavers" searched through northern New England for additional farm girls to work in the factories, and as they came back with fewer and fewer recruits, mill owners hired refugees from Europe's wars and famines, men and women willing to undertake any form of employment. Also, when the Panic of 1837

Major inventions stimulate subsidiary inventions. Here the teeth of an experimental rubber cracking machine pulverize Brazillian rubber at Goodyear Tire Co. in Akron, circa 1929.

was followed by that of 1857 (during which many mills in Lowell and Waltham failed), Lowell, once a kind of utopian dream, a pleasant money maker for privileged Bostonians, now became a harsh warning that large-scaled technology could be pushed too far.

Meanwhile, other smaller-scaled inventions were accelerating changes in the way Americans led their lives, even as those lives were swept along in the tumultuous, uncontrolled course of the national economy. These inventions included Charles Goodyear's finally successful "vulcanization" of rubber—a process which the desperate inventor, pauperized by the Panic of 1837 and humiliated by a series of failed experiments, stumbled upon in 1841 when a piece of rubber was accidentally charred, and cured, on the top his brother-in-law's stove. It also included the sewing machine, which Elias Howe patented in 1846, a decade after Walter Hunt had perfected the crucial "locked stitch" which made the process feasible.

So, by their creations, American inventors made clear that they had the power (if they persisted and if they could find sufficient backing) to revolutionize the living patterns of the emerging middle class. With the improvement of the U.S. Patent Law in 1838, inventors like the dogged experimenter Goodyear and the superior mechanic Howe (who had received his training in the Lowell machine shops) became the new heroes of American culture. Whereas only 500 patents had been granted before 1838, thousands would be catalogued and secured by the patent law in the decades to follow.

The flexibility of U.S. corporate law—based on Daniel Webster's famous Dartmouth College Case in the Supreme Court, which found the college to be an

Inventions ranged from grand to trivial. The 1846 patent application below depicts an industrial system for evaporating water. The worker at left shows off a new rubber cuff on a glove.

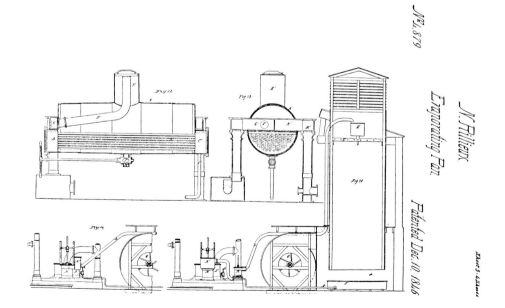

N. Rillieux's Evaporating Fan. Patent #4,879. Dec. 10, 1846.

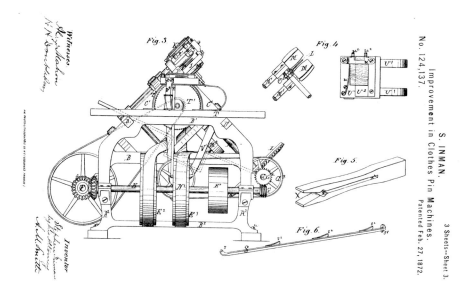

S. INMAN.
Improvement in Clothes Pin Machines.
No. 124,137. Patented Feb. 27, 1872.
3 Sheets—Sheet 3.

The inventor of the trap below imagined that an animal, passing beneath the device stuck into the soil, would trip a trigger, releasing lethal spikes. At right: a machine for improvement of clothes pins.

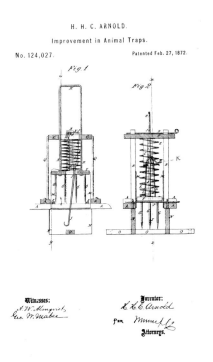

H. H. C. ARNOLD.
Improvement in Animal Traps.
No. 124,027.   Patented Feb. 27, 1872.

independent corporation, free of state interference—also favored the business of invention, granting protection to new products whenever possible. Inventors, it appeared, had the incentives and the encouragement as well as the power to solve any problem on earth—while becoming very rich in the process.

The most deeply affecting invention during the antebellum era—in terms both of people's lives and of the development of the American economy—was surely Cyrus McCormick's reaper. In the summer of 1831, young McCormick, a Virginia mechanic, showed a skeptical audience that his machine could cut its way through six acres of oats in one day. But, critics wondered, what was the use of such a clever, high-capacity gadget in the limited fields of the average farmer? They were much more impressed with the "singing plow" invented by John Deere two years later: Its circular blade seemed to clean itself of even the stickiest mud in the most difficult field.

Then came the Panic of 1837. In this economic disaster, which touched many Americans (for example, all coal mines in the commonwealth of Pennsylvania had been forced to close), Cyrus McCormick's family also suffered financial ruin. The young man now had nothing but his scorned invention to pin his future hopes on. Moving to the Midwest, he saw for the first time broad fields and open plains where his reaper could come into its own. Therefore, having also found a partner able to put up the necessary $50,000, he opened a factory in Chicago in 1847. It embodied all the principles of the American Factory System, programmed for mass production. Furthermore, McCormick discovered, in a way previously unknown to American industry, the power of advertising. To the astonishment of the world, he soon announced that his company was fulfilling orders for a thousand machines a year.

McCormick's reaper (which, as with many other inventions, was actually the culmination of numerous other reapers constructed during the same period) soon revolutionized agriculture in the new states of the American West. It gave settlers there a cash product mighty enough to rival the cotton of the South or the manufactured goods of the North. Thus blessed by its own richness and by McCormick's industrial phenomenon, the West loomed as North America's most potent prize; winning the West became recognized as the most important issue between North and South. Which states would be free and open, by means of the railroads, to further exploitation by industry? Which would remain servants of the southern economy?

As that question became more and more desperately debated during the 1850s, the business of communication asserted itself as an increasingly decisive factor. In 1843, Samuel F. B. Morse patented the telegraph. Dynamic new voices of national leadership at midcentury seized this invention as a trumpet to mobilize Americans in peace and war across the huge continental distances. How strange that this little clickety-click device, rejected at first as nothing more than an electrical toy, would lead to a whole new understanding of the power of inventions, for the nation and the world.

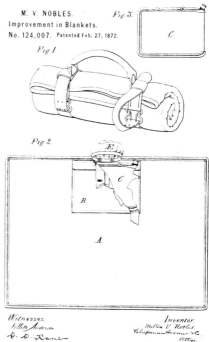

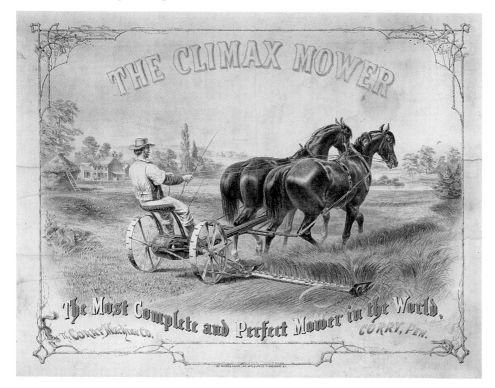

Inventions also reigned in rural America. Left: advertisement for a mower that followed McCormick's epochal machine; above: patent application for perfecting a blanket roll.

*In 1851, Samuel Colt built this showplace
factory in Hartford, proclaiming the importance
of his Patent Arms to the advancement of the
nation.*

THE ART OF INVENTION

As American inventors created more ingenious items for home and industry in the eighteenth and nineteenth centuries, and as industrialists created more innovative, mass-manufactured products, a problem that has been called The Modern Dilemma came into view. Namely, how to induce the mass of people to want more than they could decently be satisfied with, so that mills and factories could continue and increase their output.

Large-scale commercial advertising was this nation's response. Here more visibly than anywhere in the world, the industry of advertising became a giant in collaboration with the other industrial giants. Of our early inventors, Cyrus McCormick (see page 74) was the most vigorous in his use of heightened advertising techniques to insure the continuous popularity of his reapers. Other producers were not far behind: John Deere's catalogs used extraordinary language to boost their "bluebeard steel plows" and their "Gopher cultivators."

Yet it was in the extravagance of the art used to enhance these promotions that American advertising most distinguished itself. Allying themselves with new printing and graphic technologies, the artists of advertising went wild on the page, on the billboard, and wherever a consumer's eye might wander. A British visitor complained in the 1890s that the U.S.A. was daubed from one end to another "with white-paint notices of favorite articles of manufacture."

The result, of course, was that credibility was stretched to a breaking point. Inventors, already regarded as tricksters in some corners of society, became suspect as allies in the mass program of consumer seduction. Nonetheless, the ingenious and amusing ads that were left behind by those artists fill a distinctive place in the history of invention and are therefore included here in the form of a special, color portfolio.

Merry revelers salute great American inventors in this "Torchlight Procession Around the World." Among those hailed: Benjamin Franklin and Samuel F. B. Morse, along with the creator and installer of the trans-Atlantic cable in 1858, Cyrus Field and Captain Hudson. Although the seeming purpose of this lithograph was to celebrate inventors, the publisher of the ad, A. Weingaertner, actually took the opportunity to display his artistic and literary skills, promoting himself on the basis of the popularity of 19th-century inventions.

Plate 1

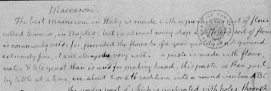
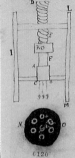

Thomas Jefferson's helpful notes accompanying his drawings of a macaroni-making machine explain to users that the plate at the top of the mill ("N-O") can have many more holes.

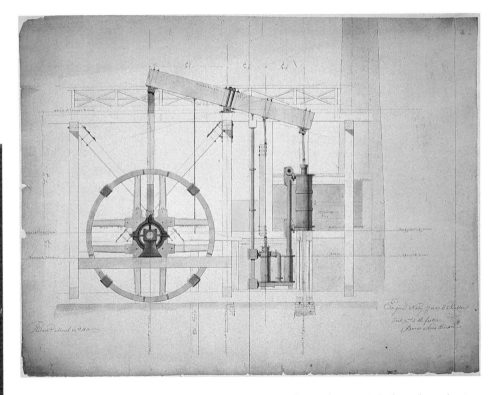

James Smallman's 1810 plan for a drive-wheel in the Navy Yard at Washington, DC, shows the application of Benjamin Henry Latrobe's low-pressure engine to a public task.

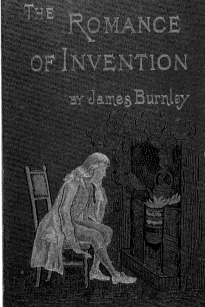

Benjamin Franklin became the promotional artist's symbol for American invention. Here his portrait adorns James Burnley's 1886 book, The Romance of Invention.

Plate 2

ADVERTISING INVENTIONS—FOR GOOD OR FOR ILL

Thomas Jefferson surely leads the list of those early American inventors who tried to ease the work of homemakers—as well as to spread the lively word of those improvements. One immediately thinks of such inventions as his macaroni machine, pictured and vividly described on the facing page. His invention of beds and desks and food-elevators also confirm the founding father's humane purpose. This is the purpose that Benjamin Franklin referred to as "inventions in assistance to common living."

There were inventors such as Benjamin Latrobe (one of whose draftsmen penned the precise plan for an engine, opposite) who announced to society their better water systems and large-scaled creations. And there were others like Sylvester Graham (1794–1851) who advertised their newly created, healthy foods—in his case, Graham flour and the crackers that bear his name—contributing additionally to the decent reputation of inventors. But there were others who demeaned that reputation by creating nothing but briskly promoted nostrums. And, for many Americans in past years, it was often difficult to tell the difference between the two species of invented products: the needed versus the merely novel.

Take, for example, the situation of William Colgate, the New York soap manufacturer whose ultimate fortune waxed large enough to endow the college that now bears his name. Colgate sincerely believed the axiom of Methodist preacher John Wesley that for healthful citizens "cleanliness is next to godliness." And that was the thrust of his promotion for Colgate products which ingeniously mixed potash from central New York with other more exotic ingredients. By dint of sweet-sounding words ("Cashmere Bouquet") and colorful ads—none so outrageous as the ad for Hamilin's Wizard Oil, below—he persuaded millions that his was the soap they needed. Was he a snake oil salesman or an inventor, or both? The pages that follow present the ads of a number of equally creative practitioners whose value may be easier to judge today than yesterday.

Medical remedies to cure real and imagined ills were produced and advertised by inventive minds as the 19th century advanced. Shown here is a Wizard Oil to cure pain "in man or beast."

Plate 3

A poor fellow being attacked by the graphically rendered demons of various kinds of pain, and facing death itself, finds joyful relief thanks to Wolcott's Instant Pain Annihilator.

Plate 4

Promising to give the patient wisdom and strength against his enemies, this ad for Old Sachem Bitters, or Wigwam Tonic, says nothing about the amount of alcohol in each bottle.

Plate 5

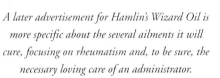

A later advertisement for Hamlin's Wizard Oil is more specific about the several ailments it will cure, focusing on rheumatism and, to be sure, the necessary loving care of an administrator.

Plate 6

More obviously an invention, the patented Universal Food Chopper was advertised in 1899 as being able to grind up just about anything on land, in air, or beneath the seas.

Plate 7

SELLING INVENTIONS FOR A BETTER LIFE

Through the newspapers and magazines which reached nation-wide mass audiences by the end of the nineteenth century, advertisers could push new inventions right into the homes of their potential consumers. Yet ads such as those on this page seem to have a certain naivete, perhaps because they pack the punch of neither modern psychology nor World War II–induced propaganda. Persuasive as they may be, as forceful as today's ads in selling a product you may not really need, they fail to blacken the lungs or excite the libido.

Indeed, the successful advertising of past inventions, great and small, may be seen as one of the major reasons why America has accomplished its historic marketing objectives, why the nation could be bound together as an economic unit, and even why Americans behave as expansively as they do. That somewhat nationalistic opinion—the view of Americans as world-class achievers—was shared by the poet Walt Whitman. In his 1865 poem, "Years of the Modern," he wrote (with advertising and invention in mind):

> *Never was average man, his soul, energetic, more like a God,*
> *Lo, how he urges and urges, leaving the masses no rest!*
> *His daring foot is on land and sea everywhere, he colonizes the Pacific, the archipelagoes,*
> *With the steamship, the electric telegraph, the newspaper, the wholesale engines of war,*
> *With these and the world-spreading factories, he links all geography, all lands. …*

These mid-1800 ads for a "home" washing machine and a "domestic" sewing machine show how U.S. inventors, initially beneficial to manufacturers, also, ultimately, assisted homekeepers.

Plate 8

THREE

AMERICAN INVENTORS AND THE NEW CENTURY

*W*alt Whitman wrote of the United States in the 1850s that it was "a nation of which the steam engine is no bad symbol." At the heart of suddenly dynamized cities, Whitman and Henry Adams and other concerned commentators discerned factories with steam-powered engines manned by workers summoned and dismissed by the blast of steam-pressured whistles. And within those hard-driving, railroad-linked industrial sites worked a number of ingenious mechanics, some of them newly arrived immigrants, some of them descendants of Yankee jacklegs. Abraham Lincoln is supposed to have joked of the latter that a typical Yankee baby, right after being taken from his mother's side, proceeded to examine his cradle to see if some "improvements" might not be made on it. To these tinkerers, even the steam engine needed improvement, or replacement.

National pride and prosperity boomed as a result of innovative installations in the industrialized cities. Indeed, after the successful war with Mexico and the hell-for-leather exploitation of the gold fields of California, it appeared that the American republic was driving to new heights not only because of steam power and steel-age technology but also because of cultural ambition. Architecturally, the clearest expression of that flag-flying aggressiveness was New York City's Crystal Palace in which was held the 1850 Exhibition of the Industry of All Nations. The central purpose of the Exhibition was to demonstrate the advanced state of American mechanical development—as well as the special brilliance of our inventors, the stars of our culture.

Planners of the much-ballyhooed Crystal Palace exhibition sought to rival the simultaneous exhibition staged in London by Prince Albert, at which American entries stole the show and the phrase "Yankee ingenuity" was heard, abroad, for the first time. Chief among the entries at both the New York and London shows were cleverly engineered, gleamingly displayed weapons, positioned in the midst of the more peaceful typewriters, sewing machines, and industrial and agricultural machines.

The prominence of the weapons was quite appropriate: In Europe, new wars and revolutions were brewing; within North America, North-South tensions were stretched

Thronging to see inventions awaiting them in the Crystal Palace built for the Exhibition of the Industry of all Nations, these Americans celebrated the materialism of their new age.

to the point of breaking. More accurate weapons, manufactured by steam-driven milling machines, would be sold to multitudes of Americans, winners and losers alike. Steam-driven railroads and electrical communication systems would also help determine both the ultimate victor and the postwar culture of the reunited nation—the nation which had been built by steam engines but which faced an electrical future.

As the post-Civil War United States flowered with telegraphs, telephones, mass printing presses, and even phonographs, British commentators spoke with some scorn of the "captive scientists" observed here, shackled in the workshops of the controlling corporations. But that was a great misinterpretation. While certain inventors at the end of the nineteenth century (epitomized by Ottmar Merganthaler, inventor of the linotype) did indeed suffer from an enslaved status even more humiliating than that endured by Paul Moody at the beginning of the century, many others did not. For them, this was the grandest time to be alive. It was a time when the American Inventor, as a superman, became even mightier than he had been in the days of Franklin or Fulton or Whitney; his options were many.

As the nineteenth century came to a close and a new century began—the "American Century," as it would come to be called—the inventor seemed positively titanic, capable of reshaping civilization with the wave of a wand. With his trumpet, he would speak across the distances, with his lightning he would illuminate the darkest valleys, with his magic box he would capture instant scenes forever, with his wings he would fly among the clouds. Corporations sought him out, beseeching him to take their money to create new products, to lend his iconic name to new items for home and workplace, to take the highest chair amid statesmen and sages at the center of the pantheon.

When the Wright Brothers became world heroes as a result of their unrivaled flights in the first decade of the twentieth century, President William Taft sent them the following message:

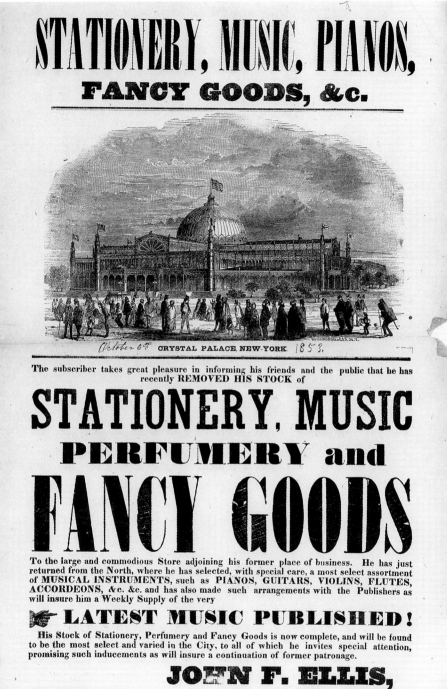

One advertisement for the Crystal Palace in New York in 1853 stressed the availability of stationery and music. It was published more swiftly thanks to advances in American printing.

The Crystal Palace's success was followed by the 1876 American Centennial Exhibition. This room featured British printing versus innovative American presses.

You made this discovery by a course that we of America feel is distinctly American, by keeping your nose right at the job until you had accomplished what you had determined to do.

Their accomplishment was distinctly American—that's what the President liked. He referred not so much to scientific genius as to untiring determination and by-guess-and-by-God daring. It may be unlikely that the President, on considering the Wright Brothers, had in mind the example of William Kelly, the hillbilly who might be called the American inventor of modern steel-making. But no one could serve his point better. This figure, Kelly, with his rusting equipment and trial-and-error techniques, deserves to stand alongside Sir Henry Bessemer, the Englishman who, by science, "discovered" how to manufacture steel in a modern setting.

Sir Henry's conclusion of 1855 that iron required a sudden infusion of oxygen to become finished steel, coupled with his laboratory discovery that additional carbon and other chemicals should also be supplied, implied among other things that, henceforth, the world would need a more learned, a more laboratory-equipped variety of scientist in order to make significant, technological advancements. Inventors must wear white smocks, like Sir Henry, it would appear. But in the little town of Eddyville, Kentucky, William Kelly, a dungaree-clad artisan of iron kettles, was figuring out at the same time how to manipulate his basic iron materials into the composition of steel. He determined by observation that "air was fuel" to the iron; only by massive blasts of it might the iron be converted into steel. His experiments with oxygen additives became so compulsive and time-consuming that he had to hide his hearth in a thicket from his father-in-law, lest the old man think he was merely fooling around and not earning serious money to support his daughter. And so Kelly, by sheer determination, came independently to the same conclusion as Sir Henry.

At first America's new industrial giants took little notice of William Kelly's discoveries, nor was his legal position honored. The divinely discontented kettle-maker had

In both the North and South the Civil War provided opportunities to try out new technologies and materials, most dramatically in the 1862 fight between ironclads Monitor and Merrimack.

the additional misfortune of being wiped out financially by the national panic of 1857. But in later decades, Kelly (who, before his death in 1888, had registered his concepts with the U.S. Patent Office) and his heirs put up a strenuous fight against the representatives of the Bessemer process. The result: a massive, notably successful industrial merger.

The backwoods steel-maker, therefore, may have been viewed very properly by President Taft as a reminder that, for all the superiority of European science in the nineteenth century, home-grown American tinkerers could continue to give the world lessons on the nature of invention. By their success, by their determination, they would make the point that, in the words of Thomas Edison, invention would always be "ten percent inspiration and ninety percent perspiration."

And in nineteenth-century America, perspiring creatively seemed to be what living was all about; as historian E. L. Bogart expressed it, "invention was a national habit." Yet there must be a limit, it seemed, to all the sweaty creativity of our culture. Just as the basic elements of water and wind and fire had begun to play less essential roles (given the new strengths of petroleum and electricity and amalgamated materials), so American ingenuity itself, along with upcountry determination, might become obsolete. At the end of the century, that's the way things looked to Charles H. Duell, President McKinley's Commissioner of the U.S. Patent Office. He concluded that now that man could drive anywhere, make anything, talk across the oceans, invention as a creative force was simply not necessary. "Everything that can be invented has been invented," he wrote, and he urged the abolishment of his own office.

The creative force of invention had roots far too deep within the American character and within American culture to be eradicated so easily. Indeed, during the nineteenth century those roots both sank deeper and produced more determinant patterns of culture than in the preceding century. The effect of invention on the next century would be even more intensive, the flowering even more abundant.

It was thanks to the determination of wondrously creative, individualistic characters such as Samuel Finley Breese Morse and Thomas Alva Edison and the Wright brothers, operating as one century ended and the new one began, that American invention did not expire or become enslaved by either industry or government. These

In these two extravagant advertisements, manufacturers hail the advanced processes that give their products primacy. At right, D. D. Badger & Others reveal the creative industriousness of their smoke-spewing plant—the forging, milling, and rolling of architectural iron. At far right, the Haish Company of Chicago goes all-out in cartoons and doggerel to assert the excellence and legality of its enameled and galvanized steel fence wire with improved "S" barbs.

As America industrialized its agricultural technology, improved plows and reapers followed the cavalry west. This deep-cutting plow was offered by North River Warehouse.

One of the most reprinted images of 19th-century America: General Custer and his surrounded troopers blaze away with Colt-made carbines and revolvers at Indian attackers.

inventors could not, of course, see all the consequences of their inventions—the mass battlefield slaughters by improved weapons, the further debasement of industrial workers—and they were as guilty as other Americans of seizing any opportunities that came their way, whatever the cost to others. But, as cultural heroes and social leaders, they did indeed light the shadowy way into our own times.

I CAN MAKE ANYTHING A BODY WANTS— THE SPIRIT OF SAM COLT

Mark Twain admitted he was in love with machines of all kinds, imagining them in his fiction as steeds to be harnessed for a drive to wealth and ease. Yet in the nonfiction, real-life world, he, as Samuel Clemens, lost a good deal of his fortune by investing in innovative printing machines and other complex machinery. Nonetheless, he like many others of his countrymen worshiped unfailingly at the altar of American industrial ingenuity. Twain focused his devotions specifically on the inventor Samuel Colt, creator of the famous Colt revolver, who struck him as master-builder of the American cosmos. In *A Connecticut Yankee in King Arthur's Court* (1889), Twain's hero is head superintendent within Colt's steam-powered factory. This proud character exults:

> I am an American. I was born and reared in Hartford, in the State of Connecticut—anyway, just over the river in the country. So I am a Yankee of the Yankees—and practical; yes, and nearly barren of sentiment, I suppose—or poetry, in other words. My father was a blacksmith, my uncle was a horse doctor, and I was both, along at first. Then I went over to the great arms factory and learned my real trade; learned all there was to it; learned to make everything: guns, revolvers, cannon, boilers, engines, all sorts of labor-saving machinery. Why, I could make anything a body wanted—anything in the world, it didn't make

As immodest as his factory, Samuel Colt's mansion in Hartford displayed all the panache of its inventor-owner. Having "made hay" during the sunshine of a frantic life, Colt died at 48.

any difference what; and if there wasn't any quick newfangled way to make a thing, I could invent one—and do it as easy as rolling off a log. I became head superintendent; had a couple of thousand men under me.

More recent generations of Americans have not found Samuel Colt to be quite the young, silvery god he seemed to Mark Twain. Colt's tarnish problem, as seen today, was not only that he did not invent much of anything but also that he was politically and socially irresponsible. This raises the question of how and why he loomed so supernally in his own day—why such a scalawag came to set the pace for other nineteenth-century inventors, as well as for those who would follow in the next century.

Born into a wealthy Connecticut merchant family in 1814, Colt was keenly aware of the fact that fortunes might be either won or lost. His father had done both in the West Indies trade. By the time he was ten, the boy had revealed himself as more of a prankster than a scholar. It seemed best, therefore, to take him out of school and put him to work in the silk mill at which his father was then trying his hand. But that employment did little to discourage young Sam, who found time to experiment with things that exploded. At fifteen, he shocked local sensibilities by inviting all to come and see how he could blow up a river raft; so tremendous was the water shower from the explosion that most of the crowd went home wringing wet. The boy himself was shielded from the force of the water only by a large, protective, suddenly intervening body which belonged to a friendly mechanic. The fellow was named Elisha K. Root; he would later nurture Colt's industrial empire with his particular genius.

Young Sam's family soon concluded that he was too "fractious" for polite society and sent him off on a sea voyage in 1830, outward bound from Boston to Calcutta. One day while on the homeward voyage Colt was strolling about the ship, dreaming about how to make faster-acting firearms. All at once, his gaze fell on a shipboard windlass. Here the myth of Colt the young inventor begins, for it was allegedly in that inspired moment that he arrived at the idea of a pistol breech rotating on its axis, just like the windlass. A popular statue still visible in Hartford depicts the clever lad carving a revolver out of a hunk of wood to demonstrate this mechanical solution. The carved pistol, later rendered in metal, became the "invention" which Colt patented in 1836 (U.S. Patent #138), the five-shot revolver on which his reputation was made.

Modern writers challenge the originality of Samuel Colt's concept. When visiting the Tower of London on his voyage, the wanderer had seen an exhibition of repeating fire arms. When in Boston he may also have seen Elisha Coller's flintlock of 1813 equipped with a rotating chamber. To give him some measure of credit, it must be added that while the chamber of Coller's musket had to be turned by hand, Colt's was turned automatically; by the action of a pawl when the revolver was cocked. But Colt certainly did not work out the pesky mechanical details of his concept. They were left to an "unnamed mechanician" who had been hired by Sam's father to produce a working model for the patent presentation.

However, the patent was in Samuel Colt's name. And, with the gifts of a natural born salesman and with every confidence in his idea, he set out to open shop and make millions. But when one of his first two completed revolvers exploded on use and the other refused to function, he admitted momentary defeat. Taking to the road, he looked around for anything else to sell, to anyone who would buy—even laughing gas in Canada.

Returning home to try again, he borrowed money from his father to sail off for a fund-raising trip to Europe. Eventually he acquired enough capital ($200,000) to start constructing a plant at Alexander Hamilton's nearly defunct industrial site of Paterson, New Jersey. He called his enterprise the Patent Arms Manufacturing Company, emphasizing his hold on *the* patent for automatic weapons. He succeeded in winning a contract for 100 repeater carbines and also produced enough five-shot revolvers to win the "Colts" eternal fame among rangers of the Texas frontier. But those successes were not enough to keep his poorly managed enterprise from collapsing. Again he was out on his ear, or, more accurately, out on the road selling and selling, increasingly determined to make his own fortune somehow, one day. It was only with the commencement of the Mexican War, when the U.S. government came through with a long-sought contract worth $28,000, that Colt could begin manufacturing and distributing weapons again.

Like Eli Whitney, when he had won his government contract back in 1794, Colt in 1845 had no plant, no work force, no production system. He did, however, understand that to be respected as an armorer (and to complete the contract), he had to utilize the "American System of Interchangeable Parts." He understood, too, that this concept, because of its unfeasibility in the early days of machine milling, was honored more in theory than in actual procedures. Colt was actually heard to boast in later years that the parts for his early revolvers "did not come close to being interchangeable!"

Maneuvering smartly, Colt turned to the originator of that famous interchangeable concept and asked Whitney himself to produce the new generation of .44 caliber six-shooters. Whitney was by then becoming a master of milling machinery, operating within the tradition of Connecticut's most famous arms maker, Simeon North. As if to prove himself, Whitney went to work on the order zealously with his new and increasingly precise machines, completing it in six months, to the rather astonished satisfaction of

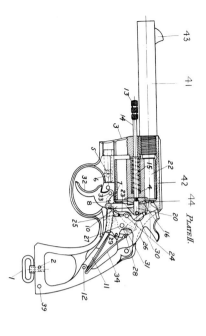

This diagram of a Colt revolver shows the coordination of trigger, hammer and cylinder that results in readiness to fire again after the first shot. The ring on the handle is for a connecting thong.

both Colt and the U.S. government. Forever after, the very name "Colt" was a byword for smoothly functioning, lethally effective sidearms. Thanks to Eli Whitney.

As he sought to expand his arms empire, Colt relied on superior machines to make the parts for his various weapons. The process seemed to work well—eventually, only twenty percent of the manufacturing in his Hartford plant was carried out by old-time artisans. And, to his operation's everlasting benefit, Colt also relied on the superior genius of Elisha K. Root. Lured away from a nearby ax factory, Root—the same genial mechanic who had sheltered an explosives-crazy kid from the water blast years before—almost nixed the arrangement by demanding the startling salary of $5,000 a year. Colt finally paid it, and was undoubtedly glad that he did. Root's brainpower was responsible for all new inventions from the Colt works; and it was he who guaranteed that the manufactured parts of any weapon would be perfect.

Within less than a decade, Colt became a millionaire. Freed at last to unleash his creative powers, Colt chose, instead, to be the never-ceasing, hard-driving sales director. Oh, he dabbled with other inventions, such as waterproof cartridges and Fulton-like submarine explosive devices (for which he received a $6,000 grant from the Navy), as well as tamper-free wire for Samuel F. B. Morse's newly invented telegraph. But his energies were consistently directed toward making maximum profits from his arms factory. In 1851 he transferred all his operations to a new showplace on the Connecticut River at Hartford, for which he engineered an impressive, two-mile-long dike. Though it was customary in New England at that time to adorn factories with cupolas and weathervanes, no one had ever seen anything like the gilded dome that Colt raised above his armory. Called "eccentric" and "ostentatious" in its day, it was designed in much the same spirit as the crystal palace: Look what an impact inventors can have on the world in which we live!

Behind this showmanship lay both high spirits and ruthlessness. Casting aside whatever ethics he may have inherited from his pious ancestors, Samuel Colt paid military officers ("under the table," in the phrase of the day) to become his no-questions-asked salesmen across the nation. He put up $50,000 as a bribe to the U.S. Congress to extend his exclusive manufacturing contract (to their credit, the Congressmen declined). He spoke with utter cynicism of his blazing "Peese Makers" and the kind of peace that they in fact brought to the frontier. These blazing guns were still the primitive six-shooters; the famous Colt .45s did not appear on the frontier until after the Civil War.

As that war began to loom as a reality—a reality long dreaded but deemed not illogical by many Americans who had feared the Union would necessarily break up into a number of republics—Colt ran his factory night and day, selling carbines to any and all who applied. As a Democrat, opposed to both the "black Republicans" and the extreme Abolitionists, Colt declined to discriminate between North and South as customers. Without a thought, he fired any workers who voted Republican.

Perhaps this unquestioning quest for profits should be regarded as expectable in that era. But for him to keep on in that pattern *after* the time of Secession, for him to disguise as HARDWARE a shipment of arms to the South three days after the bombardment of Fort Sumter, can only be viewed as unconscionable. Yet that action was characteristic of this particular man and of this new species of winner-take-all inventor, not willing to be bossed, not encumbered by social or political conscience. When asked about his attitude, Colt said that he intended to "make hay while the sun shines" as long as he could.

In fact he enjoyed that boomtime for but a few years more. Having driven himself ceaselessly to capture the dollars and the prestige which had so long evaded him, he succumbed to illness in 1862 aged 48, in the middle of the war for which he had so carelessly supplied the weaponry. Elisha Root took over as president of the Colt Armory. But Colt's explosive, aggressive spirit lived on. In the decades to come, the splendid ethos once espoused by Franklin and Lowell, according to which inventor and even capitalist should serve society, would assert itself but rarely. This shift in emphasis within the dynamics of American invention was one of the accidental accomplishments of Samuel Colt. Other contemporary inventors gave him a helping hand.

"WHAT HATH GOD WROUGHT?"

A contemporary of Samuel Colt, Samuel Morse shared with him more than an ancient New England lineage—he shared pretentiousness. Though recognized as inventor of the telegraph, Morse actually was overrated in his own time and in history for the originality of his contributions. Oddly, like Colt, he got the biggest idea of his life when returning home on shipboard—an idea given him by somebody else. Making a fortune from that idea, he left behind him a record notable for its pretenses and political reactionaryism. Yet like Colt, he too was a kind of genius, full of daring and determination; his name, and his imprint on American culture, would not be forgotten.

Born in Charlestown, Massachusetts (in the shadow of Bunker Hill) in 1791, Samuel Finley Breese Morse was the son of a distinguished preacher and geographer. Thus his playmates, even his classmates at Yale, tended to call him "Geographer." He soon found that the easiest way to earn pocket money was to produce a miniature portrait of an eager sitter at a dollar a likeness. For the more elegant members of his clientele, Morse would paint not on wood or canvas but on ivory, five dollars a portrait. Returning from a post-collegiate, three-year term of studying art in London, he moved to New York, ready to paint large and educational canvasses in the classical mode. His exhibitions, praised by critics, were honored by the local citizenry mostly through their failure to attend. Yet, sustained and encouraged by close friends, he helped found the National Academy of Design, of which he remained president for two decades. In 1829, he returned to Europe for further study.

It was when Morse was sailing home from France on the *Sully* in 1832 that his educated mind suddenly became alerted to a new and possibly valuable development. His fellow passengers were conversing excitedly about the work of André Marie Ampère, professor of Physics at Paris's Ecole Polytechnique. Ampère's experiments had demonstrated how a needle, swinging in response to an electric current, could be manipulated by an operator and could spell out a message. As he heard of these experiments (and as he recalled lectures in electricity at Yale under Benjamin Silliman), Morse's mind leaped ahead to the design of an electrical signaling system. It would span great distances, it would win him a fortune—and, yes, it would benefit the nation.

The word "telegraph" at that point generally referred to a semaphore system. But to Samuel Morse, the term came to mean an electrical signaling system. "The lightning would serve us better," he said, than any other medium for transmission. "By sparks" he would manage, somehow, to send long-distance messages—though at this time electricity was still viewed as a strange *fluid,* with only a few scientists having a dim idea of whence it came and how it behaved. Young Morse was definitely not among them.

Again he struggled to support himself by portraiture (painting, among many others, an excellent portrait of his friend from New Haven, Eli Whitney). But painters found life hard and commissions few in the newly materialistic America of the early industrial era. Though Morse won a position teaching art at the University of the City of New York, and though his affability and speaking abilities won him audiences in political circles, he could not earn even enough money for the necessities of life. Once, when the professor's emaciated condition shocked a student whom he met on the stairs, the youth asked him if ten dollars would possibly be of service. Morse replied that "ten dollars would save my life; that's all it would do." So the money was given and gratefully received.

Meanwhile Morse dreamed of his telegraph, obtaining here and there the wires and magnets he thought necessary for the type of system he had in mind; at first he was unsure of even such details as how a battery made electricity. His single, stated purpose in completing an electrical telegraph was to win enough money to afford his artistic vocation. In these early years of the 1830s, as he labored on his ungainly telegraph (which at first looked, not surprisingly, like the stretcher for a painter's canvas) he was completely removed from helpful scientific developments in the United States, in Great Britain, or on the Continent. In all of those locations major electrical discoveries were being made—while Morse was producing little more than short circuits. But, even as friends concluded that his "miserable delusions" would be the end of him, his apparatus started to respond as desired, impulses being passed along the wire by his direction. The year was 1835.

On the political front, young Samuel Morse's efforts also became more serious in mid-decade. Backed by a number of New Yorkers who respected his true-blue lineage and his professorial credentials, he entered the race for mayor in 1836, running on the so-called Nativist ticket. This peculiar party, anti-Catholic and anti- just about anything else that was not in the classical, republican image of the United States, said more about Morse's Bunker Hill background and social perspectives than about his national political philosophy. (Intellectually, he was a Democratic, opposed to the conservative Whigs.) It was quite a nasty, spiteful contest which, fortunately, Morse lost. Unabashed, he accepted orders from the party's old guard to run again in the even more brutal campaign of 1841.

He remained fortunate in other supporters, however. A fellow faculty member named Leonard Gale introduced him to the writings of Joseph Henry, the notable American scientist and theoretician of electricity who had succeeded in designing an electrical telegraph in 1831. Henry had succeeded, that is, in ringing a bell, as desired, at the far end of a wired system—an epochal step in the advancement of telegraphy. Alfred Vail, another friend, began to see the possibilities in Morse's experiments and agreed not only to construct workable models of Morse's telegraphs at the Vail family ironworks in New Jersey but also to supply funds for setting up a manageable business. Both Gale and Vail became partners in Morse's undertakings; they could see that a gamble on his work might be worth the risk.

By 1838, Samuel Morse had developed the unique system of dots and dashes known ever thereafter as the Morse Code (which idea, of itself, was not original with him). With

In this portrait of an aged S. F. B. Morse, he sits between his twin talents—a portrait and his telegraph. He was also noted for his public discourses, such as this one in the title page below.

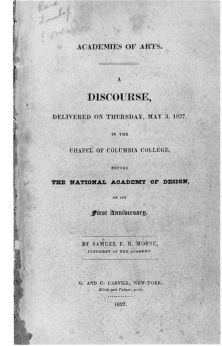

the encouragement of the prestigious Franklin Institute in Philadelphia, and after an enormously instructive meeting with Joseph Henry at Princeton, he felt prepared to request federal government support in 1838. He proposed to build a test telegraph line for the distance of a hundred miles—between, say, Philadelphia and New York. Unfortunately he made the proposal at the time of America's worst economic crisis to date, the panic that had begun in 1837. All ears were closed to such requests.

Discouraged and frustrated, Morse set off to England in search of support. But there he found, amid the flurry of Queen Victoria's coronation, a frenzy of activity on the telegraphic front. With wide-opened eyes, he saw his competition in action; he returned with little but heightened anxiety to reward him for the voyage. Yet he did succeed in securing his U.S. patent for the telegraph in 1840—eight days, as it happened, after two Britons received their patent for a notably similar device.

He then entered upon a program of beseeching Congress for financial support, a program that would do credit to the most energetic of today's lobbyists. Crucial to his promotional position was the fact that Joseph Henry, with whom he continued to correspond, had publicly stated that Morse was not merely a "mechanic" who had been able to put together a working telegraph. No, Morse was, for all practical purposes, as Henry then expressed it, the instrument's "inventor."

On the strength of that boost and of his awareness that he was now in the public eye, Morse asserted himself even more strenuously. He announced that he intended to construct a telegraph line between Washington and Baltimore, where the Whig Party convention of 1844 would be held. And he had every confidence that Congress would advance the necessary funds. Congress, in fact, did nothing but debate, no-action weeks becoming no-action months and years. Exhausted, seemingly defeated by cynics and old enemies—including J. Q. Adams, who would never forgive this son of Massachusetts for having left the Federalist Party—Morse was packing up to leave the capital one morning when he received wondrous news. The night before the Senate had passed a bill that would give him the necessary $30,000 for his proposed telegraph line. Morse had finally triumphed.

A number of technical difficulties had to be overcome before the telegraph wire could be laid along the route to Baltimore—the foremost difficulty being laying the underground pipe which would carry the wire. For the installation of this pipe a clever mechanic named Ezra Cornell had invented a plow that both dug and buried as it went along. But the improperly insulated wire refused to function. Finally, with the wire strung on poles (according to a design worked out by the same Cornell, who used broken bottle bottoms for insulators), the system was successfully installed along the forty-mile route to Baltimore.

Well known in fact and legend are the first words transmitted: "What hath God wrought?"—a message made up not by Morse but by the daughter of Morse's ally, the Commissioner of Patents. Less well known is the first useful job accomplished by the telegraph: Morse transmitted vivid reports to his Congressional audience about the unexpected emergence of James Polk over Martin Van Buren as the most likely Presidential candidate at the Baltimore convention. Cheers for Morse, *the inventor,* rang in Washington even more loudly than cheers for Polk. Cheers, at last.

The press hailed Morse as the nation's new hero; a Utica, New York, newspaper opined that his telegraph was "unquestionably the greatest invention of the Age." In

A sentimental Currier and Ives print recognized great American inventions of the 19th century: the lightning steam press, Morse's electric telegraph, the locomotive, and the steamboat.

the Mexican War of 1845 it proved its value for long-distance communications. Yet, in a crucial decision, the government declined to acquire the telegraph as a national utility—an action which had been taken by the Republic of France, among other nations. But as official interest in the telegraph dwindled, a new breed of utility capitalists (including Cornell) moved in to make millions from Morse's Magnetic Telegraph Company and the subsequent Western Union Company. Happily for Morse, various court rulings, leading to the Supreme Court decision of 1854, determined that the telegraph was entirely his invention, deserving of royalties from all users.

Unhappily for Morse's personal reputation, however, he chose in the course of achieving that recognition to deny the substantial help given to him at a critical point by Joseph Henry (who had become the first Secretary of the Smithsonian Institution). He also neglected to credit the roles played by Gale and Vail (who had done away with Morse's original "composing stick" and invented the balanced key by which the telegraph would be operated). He had seen a chance to be called the one and only inventor, he had grabbed it, and the nation's corporation-favoring legal system had supported him. Today's historians tend to believe otherwise, identifying Joseph Henry, the shy and retiring scientist, as the only one on the scene at the time with sufficient knowledge to have completed a proper electrical telegraph.

There is no doubt, however, that Morse applied himself to the technological challenges of his day with an outstanding intelligence and an unusual determination. No one denies him that. For example, when Morse met the artist and photography pioneer Louis Daguerre in France, the Frenchman was much impressed by the American's quick understanding of the technical processes. They agreed that Morse should open a studio upon his return for the making of daguerreotype portraits. And, in operating that studio, Morse made so many improvements in the process—before the expense and the time required for the portraits shut it down—that in many accounts he is hailed as the father of American photography.

What was Samuel Finley Breese Morse, then? An accomplished painter and a determined experimenter, a fierce promoter and builder who showed all who could understand that the American political/financial structure would eventually support new ideas in communications. To would-be inventors of his and later generations, he became a symbol of what the amateur technologist could then do—he could, for example, revolutionize world communications and make a fortune.

Morse spent his final days at his estate, "Locust Grove," not creating grand art works but dispensing philanthropic contributions and no-longer applicable political wisdom. Opposed to the Civil War, he joined the Copperhead wing of the Democratic party. And, until the day of his death in 1872, he would not be shaken from either his nativistic credo or his pretentious belief that he alone had invented the telegraph.

"THE EIGHTH WONDER OF THE WORLD"— THE LINOTYPE

It was Thomas Alva Edison who used this lofty phrase, "the eighth wonder," to assert the importance of the newly created linotype machine, an invention that came to prominence soon after the perfection of the telegraph. Though probably not the eighth or even tenth world wonder, this was the historic machine of 1884 which finally took typesetting out of the hands of medieval-style typesetters and made it into a modern, mechanical process (the machine driven by steam). Others joined Edison in regarding it as of prime importance in the advancement of culture, an even more essential step than the telegraph in assisting communications among the mass of Americans. They were fond of saying that "not since Gutenberg's Bible" had there been such a breakthrough, such an advance toward literacy for all mankind.

Who, then, was the inventor of this wonderful machine? Named Ottmar Mergenthaler, he was born in Württemberg, Germany in 1854. After being trained as a watchmaker (like so many inventors before him), he came to the United States at the age of eighteen with every expectation of making a good living as an instrument or model maker. Learning from a cousin that the U.S. Patent Office was located in Washington, D.C., he first headed there for an appropriate job. But, when his cousin's firm moved to Baltimore, he went along. In that bustling port he soon found himself caught up in local debates about how to improve the typewriter and, along with it, the printing machine. If Samuel Morse could make a fortune in the previously unknown realm of telegraphy, could not a clever fellow be even more confident of succeeding in the known business of newspaper printing?

Steam-age America had already made two major contributions to the advancement of modern printing: Isaac Adams's flat bed and platen presses of 1830 and 1836; and Richard Hoe's rotary sheet press, which had been put to work on New York newspapers in 1848. Before these, in 1822, Vermont's William Church had invented a composing machine which greatly speeded up the operations of the pressman. Also in midcentury America the typewriter had its genesis, in a machine designed and patented by a Wisconsin printer named C. L. Sholes in 1867.

The exclusive purpose for the typewriter, initially, was as an aid to commercial printers. Not until the end of the century, when Philo Remington saw possibilities for the machine as a mass-consumption product, did it become a successfully established tool for office and home uses. The hope of Mergenthaler and his associates (including the brilliant Charles J. Moore, who had created a speed-typing system) was to unite the keyboard with a type-composing machine. But as they—along with many other contemporaries—attempted to do that, they ran into so many awkward mechanical problems that a solution seemed to be forever beyond them.

Mergenthaler described the creative process which enabled him to make his breakthrough invention this way: The idea came to him in a single flash when he was riding the train one day back to Washington. He suddenly saw how his machine could set letters all in one line via matrices, could cast that single line in lead, and then could return each matrix to its proper position. When, in 1884, the machine was demonstrated to its financial backer, Whitelaw Reid of the New York *Herald Tribune,* Reid exclaimed, "Ottmar! You've done it, you've cast a line o' type!" Hence the name linotype.

By 1886, Reid had installed twelve linotype machines in his press room. He had paid Mergenthaler $300,000 for his invention—thought to be the largest amount ever paid to an inventor. Furthermore, having helped Mergenthaler become a shareholder in the linotype sales corporation with a personal loan, Reid went on to speculate that, on the basis of the anticipated riches, the inventor would soon be "returning to his fatherland a rich man, occupying with his family one of the noted and beautiful castles on the Rhine." It looked as if here, finally, was an inventor who would not have to sweat and starve his way through years of grief before receiving his proper rewards.

It certainly appeared that way to Mergenthaler himself. Yet he attempted to treat his success modestly—as but another evidence of America's burgeoning ingenuity. Addressing the stockholders, he speechified in the nineteenth-century manner.

> I say you have honored your country, for everyone will know that this invention
> has been originated [sic] in the land which gave birth to the telegraph, the Hoe
> press, and the reaper; everyone will know that it came from the United States,
> though comparatively few will know the name of the inventor.

Almost immediately after this honeymoon, however, relations between Mergenthaler and Reid began to turn sour. In his autobiography, the inventor spoke bitterly of the false ethics that had come to characterize American business, writing that "promises made by companies are only made to be broken." At issue was not only Whitelaw Reid's increasingly strong-armed efforts to take total charge of the Mergenthaler Printing Company but also the inventor's attempts to slow down production so that necessary adjustments and additional improvements could be made in the machines.

The fight continued as the company matured under new, and increasingly uncreative, managements. Finally the attempt was made to remove "Mergenthaler" from the company's letterhead—in the alleged interest of increasing sales by simplifying the company's funny-sounding name. In a letter to the new president, the inventor replied eloquently:

> Of the many communications received from yourself … I do not remember
> one which has made so painful an impression upon me … . The reasons ad-
> vanced are, putting it mildly, whimsical. To deprive a man who has given to the
> world one of the most important inventions of the age of the credit therefor by

discontinuing his name, seems to me to be an act unworthy of the stockholders
… . From an original investment of no more than one and one-half millions,
the company has prospered until now it is proposed to pay interest on ten
millions … .

Rather than a castle on the Rhine, Mergenthaler's ultimate reward was total discouragement and a series of ultimately fatal diseases. He died of tuberculosis in Baltimore in 1899. Like Samuel Colt, he failed to reach the age of 50. But, unlike Colt, this true inventor was disserved by the spirit of the new age. Perhaps the moral of his tragic life is that, in the mode of both Colt and Morse, nineteenth-century inventors had to be as sharp, as occasionally deceitful, as the corporations and other agencies with which they contended for what de Tocqueville had called their proper "wealth and fame." This sharp edge, this aspect of American ingenuity seemed difficult for the brilliant, disappointed German to acquire.

EDISON, THE PRINCE OF LIGHTNESS

To Thomas Alva Edison, acquiring that sharp edge was as natural as ejecting tobacco juice into a spittoon. Though he had the sensitive hands of an artist—and , like Morse and Fulton, could very handily sketch whatever three-dimensional object was on his mind—he had the rough deportment and rude work habits of America's nineteenth-century shop culture. He could never, would never regard an invention as a scientific abstraction; it was a thoroughly practical, commercial matter. In a day of prosperity he did, in surprising fact, buy up some of S. F. B. Morse's drawings. But that was also about the time when he voiced the personal opinion that ethics were incompatible with business—and that he, as an inventor, was *in business*.

In the years before fame allowed him to deliver himself of such seasoned and much-quoted observations, Thomas Alva Edison appeared to most friends and acquaintances to be nothing much more than a tramp operator, a "happy hooligan," swinging from one telegraphing job to another. This was the nomadic life he had grown into after boyhood years working as a peddler of newspapers on the Grand Trunk Railroad line out of his Port Huron, Michigan, home. He had been born in Milan, Ohio, in 1847 to a family that then drifted farther west, on to resource-rich Michigan. One of his early memories was of an inventive, experimental neighbor, a Mr. Winchester, who had been interested in the problem of launching a man-carrying balloon. The balloon finally did take off, bearing Mr. Winchester out across Lake Erie, never to be seen again.

Edison remembered Mr. Winchester more as a hero than as an illustration of the danger-side of experimentations. Though subjected to formal education but briefly in Port Huron, young Al soon demonstrated his curious mind. He tried to see if he could repeat some of Morse's electrical tricks; he fooled around with mixtures of various chemicals (starting a fire on board the railroad car in which he was riding at the time). Wailing railroad whistles and clacking telegraph keys and screaming black headlines comprised the exciting world in which he lived.

The Civil War came along, with many attractive-sounding jobs for telegraphers, jobs which he pursued with only one or two dollars in his pocket and with the floor his usual mattress. And thus he arrived in Boston, where he found himself in the midst of a number of other restless, scientifically inclined minds of this new age of electrical power and mass communications. Despite his lack of education, he was accepted as

Usually depicted as a roughneck and wearing clothes in which he had slept, Thomas Alva Edison, in this 1904 portrait, gives a well-kept impression (as does his topsy-turvy laboratory).

one of the Hub City's creative fraternity (which included the young speech teacher and telephone inventor, Alexander Graham Bell). In 1869, having concentrated on improving transmitters and receivers for telegraphs and having glimpsed the possibilities of developing other new products for the busy world around him, Edison announced that he would stop bumming about. From now on he would devote himself full time to inventions (specifically to the "quadruplex system" for messages simultaneously over the telegraph wires)—though he recognized that the life of an inventor might be a "stony road."

Edison's first patented invention was an electrical vote-recording device, an antecedent of the stock-ticker inventions from which he derived his first satisfactory revenues. But in Boston he could find no investors, no takers. Though perhaps the most intellectually creative city of the country in the era after the Civil War, and though regarded as the cradle of America's Industrial Revolution, Boston was no longer the financially adventurous community it had once been. Edison, asking himself what was the point of producing inventions "unneeded" by a given society, concluded that New York City (where commerce rather than science called the tune) was the right place for him.

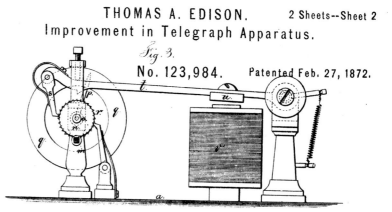

THOMAS A. EDISON.   2 Sheets--Sheet 2
Improvement in Telegraph Apparatus.
No. 123,984.   Patented Feb. 27, 1872.

Edison grew up as something of a telegraph bum, moving from job to job via the railroads; yet he was skilled enough in system mechanics to make the suggested improvement (diagrammed at right).

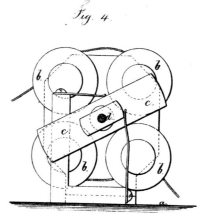

On the freight elevator's door of the main building at Edison Laboratories in West Orange, NJ, a sign reads: "For Mr. Edison's Personal Use Only." But, even when he was older, he walked up.

And it was in New York, on the basis of his successfully marketed stock-tickers, that his reputation grew. Offered a job by Western Union (which had become rightly known as the "world's most profitable enterprise")—a job which he declined in the name of independence—he began to style himself, though only in his mid-twenties, as "Mr. Edison the Electrician." Western Union, in the august person of General Marshall Lefferts, continued to regard him highly, despite his turndown of their offer. The general not only referred to Edison as a "genius" but gave him the best kind of American recommendation, stating that he was "a very fiend for work." He both helped finance Edison's first laboratory in Newark, New Jersey, and sent along a business manager named William Unger to manage his affairs.

This step seemed advisable since Edison had spent nearly all the money initially made available to him on equipment, leaving no funds for running the operation. Edison, naturally, staffed his laboratory with other young men like himself, men not with engineering degrees but with "light fingers" to detect the bugs in whatever devices needed improving. With their help, he was determined to perfect the "rapid and cheap development of an invention." In order to bring in some income to make up for that time devoted to pure experimentation, Edison sporadically demanded that his shop work around the clock to turn out the Universal Stock Printers which continued to sell handsomely.

Both the Newark shop and the others that followed it at Menlo Park and West Orange were very unorthodox work and hacking-around places, the likes of which had not been seen before. Edison vowed, when opening up the Menlo Park laboratory—which has been called the first industrial research lab in America—that he would "produce a minor invention every ten days and a big thing every six months or so." Horse-play and practical jokes alternated with through-the-night drudgery and occasional breakthroughs; hard-won successes were saluted by rowdy singsongs around a large

In an advertisement for a Phonographic Festival at which Edison's "wonderful talking invention" was demonstrated, the man himself appears as a rumpled, pensive genius.

organ installed at the back of the lab. Some of the men, having sweated alongside the master at his workbench in the ceaseless attempt to "improve odd devices and articles" like typewriters and batteries, went on elsewhere to become recognized inventors. Edison, though appreciative of his helpers, had a habit of paying them as little, as infrequently as possible.

The record of inventions turned out here in a short period of time was truly amazing—in the face of a host of competitors. Recognizing that Alexander Graham Bell's telephone had been the sensation of the Bicentennial Exposition in Philadelphia, Edison went him one better and, in 1877, made that somewhat unpredictable, magnet-controlled machine into a useful instrument by adding the carbon-button (or, as he called it "plumbago") telephone transmitter. In the same year he patented the phonograph or "speaking machine."

A modern photograph shows the early Edison talking machine, operated by a hand crank, with two victrola records; in the background are the newest versions of this technology—compact disks.

To play it, one cranked a cylinder covered with tinfoil by hand and a scratchy voice came forth. The strange apparatus survived all tests demanded of it by skeptical listeners, many of whom thought the awesome human-voice-reproducer must be a trick performed by some sort of ventriloquist.

Ironically, despite Edison's conviction that all of his inventions must be industrially relevant and commercially viable, he at first had no real idea what a good telephone or a superior phonograph could be used for. But then in 1879, understanding the marketing opportunities in an awakened America, he produced the incandescent lamp, for which gas lights had prepared the way in so many homes. To introduce the lamp to that market, Edison and his team had to create a myriad of other unheard-of devices large and small, such as generators, switches

Edison, soon before his death in 1931, is pictured here with Henry Ford in front of a brace of mikes. Ford admired Edison as a genuine American and strove to preserve his laboratories.

and sockets, and safety-fuse boxes—devices which were soon taken for granted as having always existed.

Along with these so swiftly accomplished creations, Edison (who was then hailed as the nation's greatest inventor and dubbed by America's adoring press the "Wizard of Menlo Park") had created, in himself, a new kind of inventor—a mighty force who would be a determinant of American culture. Rather than casting himself in the traditional role of the hapless tinkerer forever in search of a patron, like the suicidal John Fitch or the mendicant Charles Goodyear, he had commanded the capitalists come to him. And, assuming that a contract could be arranged for some new device, he would accept the investors' money up front. Then he would work like the very devil to produce the item.

Sometimes, as in the case of the phonograph, the result was swift and unexpected and unknown but tremendously promising. And sometimes, as with the filament for the light bulb, the complete solution to a seemingly minor technical problem was hard to come upon, embarrassingly long in resolution. Carbonized cotton thread was the final answer for the filament, a solution arrived at after more than $40,000 had been spent in fruitless pursuit of metal filaments. In such cases of discouragement and delay, "Professor Edison" was inclined to issue press releases stating that the answer was "now at hand." When he was attempting to improve the telephone receiver, the disappointments in the lab were particularly keen; the frustrations of simultaneously dealing with public opinion, rival inventors, and corporate giants reached nerve-shattering intensity. An admiring nation wondered, Would the man ever crack?

Edison occasionally succumbed to bouts of sickness and exhaustion. But usually he functioned without letup, hiding his real feelings behind the mask of a crusty, lusty, hard-driving Midwesterner who could beat the Manhattan industrialists at their own game. He associated happily with the robber barons of the Gilded Age, wrestling with Jay Gould and other connivers for the establishment of his own empire. In one contest for control of the stock-tickers, he out-maneuvered the Vanderbilt clan—leading headline writers to say that he had "bested New York's most powerful family." The crowning achievement in his drive to establish himself as a mogul among other moguls was his successful building of the world's first electric light power plant at Pearl Street in New York City (1881–1882). This direct-current plant and his sensational display at the 1881 Paris Lighting Exposition—accompanied by huzzahs! wherever he went—demonstrated that Thomas Alva Edison had become not only the world's most recognized inventor but a corporate guiding star as well.

Edison explained that he had no choice but to forsake science and technology in order to become a successful player on the fields of global industry. His acquiring of the sharp edge needed for business—particularly in the new field of public utilities—was a totally self-conscious act. "I'm going to be a businessman," he announced. "I'm a regular contractor for electric lighting plants and I'm going to take a long vacation in the matter of invention." Like another millionaire of that rough-and-tumble day, Andrew Carnegie (who is alleged to have said "pioneering don't pay!"), Edison seemed to agree that business, not invention was where the fortunes lay.

But, in fact, the real man within this self-announced baron of utilities could not stop inventing. Near the end of his life he confessed that "I always wanted money [in order] to

Henry Ford is remembered as an outstanding American innovator for his factory's high-quality assembly line. Yet not all Fords survived—for example the broken-down model at left.

Edison's invention, the Kinetoscope, took rapid snaps of moving objects, as seen in this strip of "the modern Hercules." Popular in its day, the Kinetoscope preceded motion pictures.

go on inventing." By then, however, he had demonstrated to all the world that in America the ability to stand toe-to-toe with capitalists could be an integral part of the native inventive genius. Inventors, though they had not created this new, electrified nation all by themselves, were nonetheless active partners in producing the so-called Age of Progress.

As the old century ended and a new one began, some of the inventions that continued to roll forth from Edison's laboratory—or, as we might say today, his industrial park—were outstandingly successful. These included his improved storage batteries and his motor-driven phonograph with wax records (a solution arrived at some ten years after the original invention). Other experiments helped inspire later inventors and developers, particularly his "kinetoscope" or peep show, which demonstrated how successive images and accompanying sound might be synchronized for the joint and realistic effect of moving pictures. Yet other experiments and searches led nowhere— most notably, Edison's attempts to mine and process magnetic ore, an effort into which he sank a portion of his fortune.

Edison lived well into the century he had helped fashion, dying in 1931. Unreformed in lifestyle or work habits, he continued to experiment, making real contributions to national defense during World War I in the development of chemicals previously imported from abroad. But he was by this time both a cultural adornment and something of an anachronism, not only in his rumpled suit and spittoon manners, but in his one-man-show approach to invention. The shop culture of the nineteenth century had become, by the 1930s, a not particularly fond memory. The so-called Second Industrial Revolution, in which electricity and petroleum fuels had replaced steam and water power, was well on its way to maturity—shortly to be replaced by machines driven by even more exotic fuels. And Thomas Alva Edison, who had surpassed all other inventors (with over 1,300 U.S. and foreign patents to his credit), had come to represent not the daring future but a questionable past.

Recognizing the tendency of Americans to forget their heroes, Henry Ford (who was quoted as saying that "History is bunk!" but certainly did not mean it) made strenuous efforts to preserve Edison's work place even before the inventor died. Today one may visit a recreation of his famous, cluttered, ramshackle New Jersey laboratory in Ford's Greenfield Village, Michigan—one small-town American genius having seen the importance of keeping in view the inspirational small-townness of the other.

"WATSON, COME HERE—I WANT YOU!"

Just as Ottmar Mergenthaler could not acquire the sharp, ruthless edge needed to succeed in America's business world, so Alexander Graham Bell remained something of a stranger here. Born in 1847 in Edinburgh (within three weeks of Edison's birth date, as it happened), Aleck was the son of a moderately well-known Scottish teacher of the deaf. After being educated at the University of Edinburgh and University College, London, he joined his father in promoting an innovative language for the deaf called Visible Speech. When his parents moved to Canada (following the deaths of Aleck's brothers, perhaps as a result of Britain's already polluted air), he rather reluctantly joined them in North America in 1870.

Having moved on to Boston, Massachusetts, for a teaching job in that scholarly capital, the tall and handsomely bearded young man soon established his own school of "vocal physiology." He also lectured at Boston University and traveled through the eastern United States spreading the gospel of his father's techniques. Yet he had the compulsive habit of working late through the night on various experiments. Most of these were connected with electricity, for like many alert young men in science-focused Boston, he believed that he could improve on S.F.B. Morse's telegraphic systems—for quite different purposes.

Alexander Graham Bell's purposes were almost exclusively related to the mechanical enhancement of speaking and hearing. Where other would-be inventors of the day might have been looking at the nascent industry of communications, dreaming of making a fortune from it, Aleck Bell focused on the sending and receiving of word messages mostly because he was fascinated with sound. He also hoped to assist the sound-handicapped. Years before, in his youth in Scotland, he and one of his brothers had invented a kind of wheat-husk-stripping machine, with the thought of becoming millionaires on the strength of such an agricultural improvement. But, while that youthful brainstorm had not worked out, Aleck was left with an awareness that he had certain mechanical skills—perhaps the skill to facilitate speaking and hearing, both for social benefit and for profit.

Aleck called his first adult invention—which involved wiring-up a tuning fork—a "musical telegraph." The term "telephone" had already been created, back in the eighteenth century, by German scientists to describe an imaginary instrument that might allow one to speak across the distances. As the nineteenth century matured, no one knew anything about the behavior of either sound waves passing through the air or frequencies sent over electric wires. Young Aleck, however, was an expert in acoustics, as well as a musician and an omnivorous reader in the physics of speech. After weeks and months of experimentation in realms beyond his tuning forks, he believed that he and his assistant, Thomas A. Watson, had happened upon the scientific phenomenon that would make a telephone work.

When studying the sounds transmitted between two rooms by wires connecting a primitive transmitter and receiver, Bell had discovered that a continuous electrical current could be made to vary in intensity just as sound waves vary. It was that discovery that prompted him to call out to Watson—a call which was, in effect, the world's first telephone message. Excited, but recognizing that much more work needed to be done on the instrument, the financially hard-pressed Bell looked about for someone, anyone who might provide the necessary backing. But then, would he himself be shrewd enough to handle this property in the sharp-edged manner of the new American inventor? Also, would he be able to develop the improvement of the invention on his own terms?

Luckier than either Mergenthaler or Edison, Aleck Bell swiftly found the money and the scope he needed. They were given by a brusque, mutton-chopped Boston lawyer named Gardiner Greene Hubbard (who, having been involved in the notorious Crédit Mobilier scandal, was thought by some to possess only "marginal" morals) and from a number of his entrepreneurial associates. Hubbard, originally cut from a rougher cloth than Aleck but now rich and therefore socially acceptable, shared with Aleck an interest in the transmission of sound. That interest, while having stemmed from the fact that his daughter, Mabel, had lost her hearing as a result of scarlet fever, had grown into a commercial passion—he saw sound transmission as an even better financial bet than railroads.

As might have happened in a fairy tale, the fair Mabel and the Byronesque Aleck fell in love soon after contact with the family began. But although Gardiner Hubbard

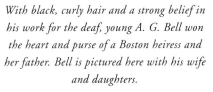

With black, curly hair and a strong belief in his work for the deaf, young A. G. Bell won the heart and purse of a Boston heiress and her father. Bell is pictured here with his wife and daughters.

was pleased enough that a possibly acceptable suitor for his daughter had appeared on the scene, he was more interested in the device that the young man was working on. As the story was told in later years by Thomas Edison, Hubbard virtually "made the successful development of an [acoustical telegraph] a condition for approving marriage to his daughter." It was an excellent connection, a perfect marriage of love and mutual interests. Although Aleck at first resisted the financial backing and imperious direction of his father-in-law-to-be (preferring, as a British citizen, to try for funds from England, or even Canada), he eventually succumbed.

Having taken an "oath of intention" to become an American citizen at the end of 1874, Alexander Graham Bell, prodded by Hubbard, received his U.S. patent (which has been called the "most valuable patent ever issued) on March 7, 1876. Between those two dates, the twenty-eight-year-old Bell made a crucially important trip to Washington, not only to apply for a patent but also to relieve himself of residual, inner fears about his cranky apparatus—which seemed to defy improvement, at least by him.

This entry for March 10, 1876 appears in Alexander Graham Bell's journals in the Library of Congress. It describes his experiments with basic elements of the first telephone.

Fortunately, in Washington, Aleck was able to arrange a meeting with Joseph Henry. On hearing of Bell's progress with variations of a continuous current, and after surveying the young man's arrangement of wires and batteries, Henry (who had himself made no contributions in this area of technology) advised, "Under no circumstances should you think of giving up!" Bell returned to Boston with the knowledge that, though only an amateur in electrical matters, his attention to the nature of sound might bring him success. It all worked: Three months after receiving his patent, Bell found himself standing tall at the Philadelphia Centennial, receiving the congratulations of international dignitaries on an instrument that successfully transmitted the human voice. It worked across the distance of the Exposition building; perhaps it would span the world.

As he demonstrated the phone across the United States—always preferring to shout "Hoy!" or "Ahoy" into the phone, rather than giving the new "Hello" greeting—his presentations benefited from his magnetic stage presence, his sonorous voice, his well-trained showmanship. He would need all those talents (plus some of the world's most gifted lawyers) to win his cases against other competitors, a host of whom instantly appeared. Through these contested years, he consistently managed to keep his dignity and health, despite punishing headaches and other neurotic ailments. His skin was a bit thin. During the tense years before his ultimate court victory, he remarked self-pityingly, "The more fame a man gets for an invention, the more does he become a target for the world to shoot at." One doubts that he would have been able to stand up any better than Mergenthaler to the harassments of a Whitelaw Reid; Gardiner Hubbard saved him from such a test.

His was certainly a different world from Edison's. Whereas the Wizard of Menlo Park concentrated on a funhouse full of technological products, determined to bang one or another of them into shape, Bell focused completely and entirely on improving his sound machine. And whereas Edison strode forth, bare-chested as it were, to do battle with the industrial titans of his day, Bell could let Hubbard and a host of lawyers (led by the wealthy Thomas Sanders) defend his cause and set up the Bell Telephone Company in 1877.

2 Sheets—Sheet 2.

A. G. BELL.
TELEGRAPHY.

No. 174,465.   Patented March 7, 1876.

Fig 6.

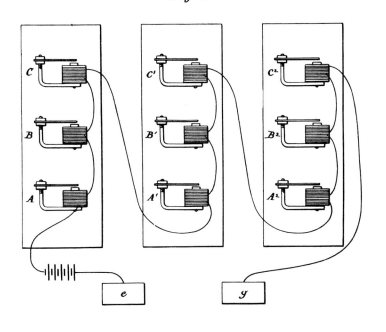

Historians call No. 174,465 at left the "most valuable patent ever issued"; it was for a speaker and sounder connected by a continuous current. A model of Bell's instrument is pictured below.

Fig. 7

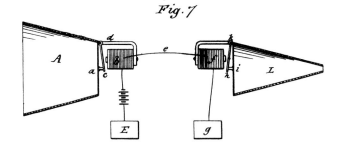

Witnesses

Ewell H. Sick
W. J. Hutchinson

Inventor:

a. Graham Bell
by attorney Pollok & Bailey

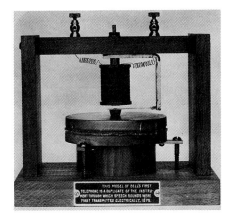

By the early years of the 20th century, when this idealized view was published, the phone operator was a national goddess, linking the land together (while wires covered the skies).

In the subsequent struggles to defend Bell's patent—which was ultimately secured by a U.S. Supreme Court ruling—the prime opponent was an aggressive corporation of mysterious membership. In that corrupt era, with Grover Cleveland in the White House, many illustrious politicos, including the Attorney General, considered it quite *comme il faut* to take part in a little corporate skullduggery. One such powerful interest mounted this most impressive anti-Bell maneuver. But Hubbard's team triumphed, even after a head-to-head encounter with Western Union, which had hired Edison to produce rival instruments. When the corporate dust settled in 1879 (when there were about one thousand operating phones in the United States), the value of each share of individual stock soared to the value of $995. The thirty-two-year-old former teacher of the deaf, who owned 1,507 shares, was a very wealthy young man, even before his company began its real growth.

Other critical differences between Bell and Edison (and other American inventors of the era) immediately became apparent. Rather than keeping the money for himself or for his laboratory endeavors, Bell gave all but ten of his shares to his wife—a totally sincere, totally sentimental act. Her advisors worked wonders with the investment, making her one of the nation's wealthiest women—even though economic historians conclude that the American telephone industries made no truly great private fortunes.

As for Alexander Graham Bell himself, he continued to pursue a course so different from the rest of American inventors that he stands as an exception to whatever rules may have been established. (Perhaps it should be noted here that Bell became an American citizen only in 1882, a step he took not hastily.) In a most un-American way, he took almost no interest in the ways and means of developing the Bell Company. Laymen of his day found this abdication almost as hard to believe as they found the basic concept of the telephone to understand.

The following conversation between a train conductor and Bell, one time when he and his wife were riding through North Carolina, illustrates in a small way the point of his turning aside from public activities. The conductor, having heard that the inventor

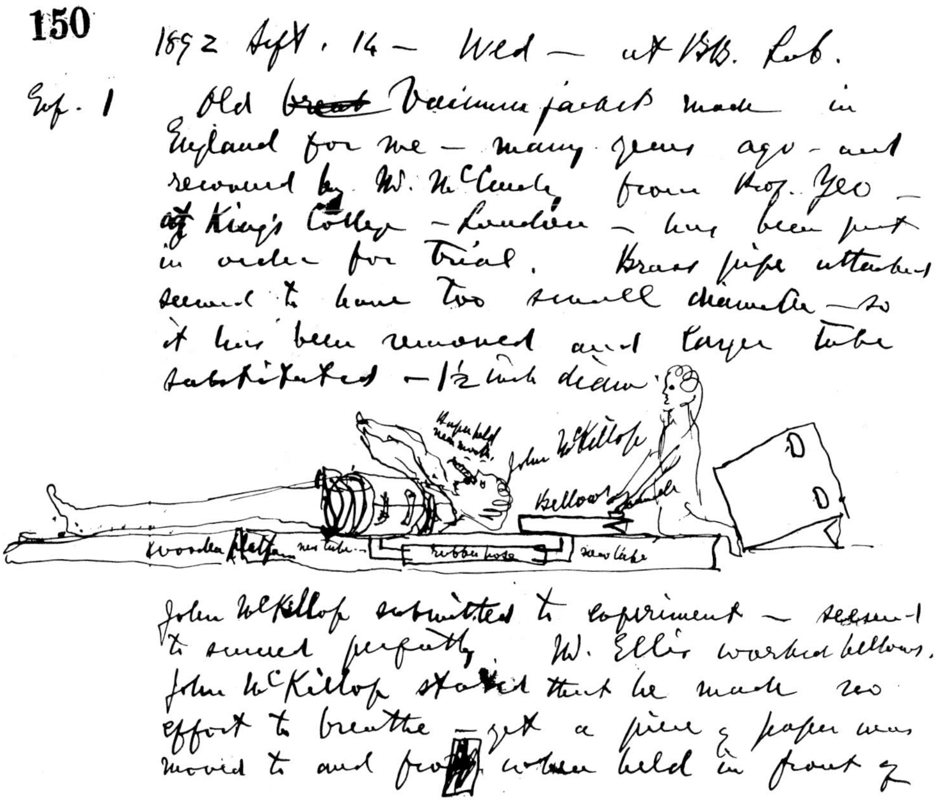

150

1892 Sept. 14 — Wed — at B.B. Lab.

Exp. 1 Old ~~breat~~ Vacuum jacket made in England for me — many years ago — and recovered by Mr. McCurdy from Prof. Yeo — of King's College — London — has been put in order for trial. Brass pipe attached seemed to have too small diameter — so it has been removed and larger tube substituted — 1½ inch diam.

John McKillop submitted to experiment — seemed to succeed perfectly. Mr. Ellis worked bellows. John McKillop stated that he made no effort to breathe — yet a piece of paper was moved to and fro when held in front of

Bell hoped to aid society with inventions: at left are his notes for a vacuum jacket to help a man breathe; the sketch below depicts his attempt to electrically locate the assassin's bullet in President Garfield.

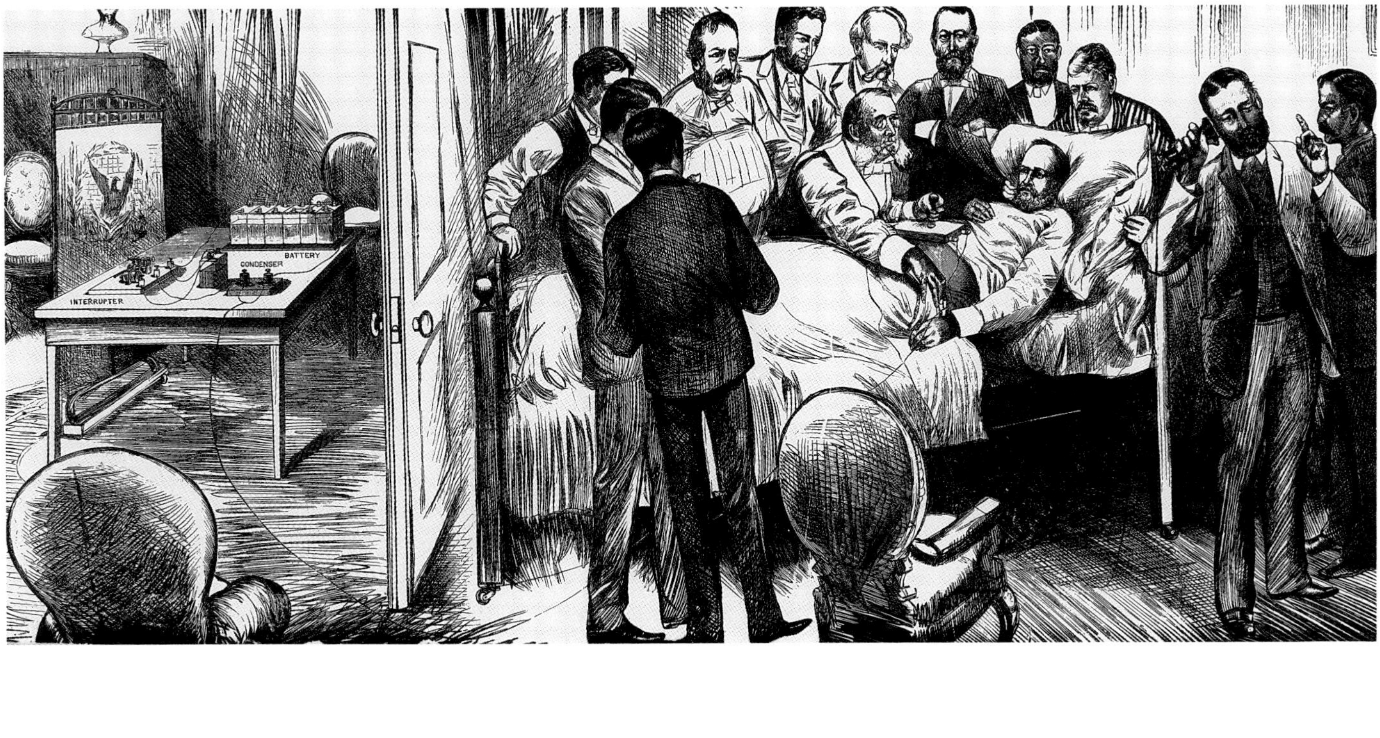

Bell explored the unheard-of idea that sound might be conveyed via a laserlike ray of reflected sunlight (as in the drawing at right above) or by means of a receiver dish (below).

of the mysterious speaking machine was on board, approached Bell soon after the train had lurched to an unannounced stop.

"Are you the inventor of the telephone, Sir?" the conductor asked.

"Yes," replied Bell simply.

"Well, do you happen to have a telephone about you? The engine has broken down and we are twelve miles from the nearest station."

All on board were disappointed when the inventor did nothing about their communications problem. In other, more important ways, he made evident his disinterest in bigger, better, faster communications systems . He was by no means lax, however, when it came to additional inventions to help the deaf, including the "audiometer"— as well as the wax phonograph record, which he developed at his Volta Laboratory in Washington, D.C. Indeed, back in the days of Edison's invention of the primitive phonograph, Bell (in a rather uncharacteristic, competitive manner) had exclaimed that, unfortunately, he had just "let that invention slip through my fingers."

Because he had accomplished so much, Bell could not stay out of the public eye. At the time of a national emergency, when President James A. Garfield was shot in the spine by an assassin in 1881, the famed inventor was called upon to find the life-threatening bullet. Earlier, he had developed a special electrical device that would cause a click in a telephone receiver if a probe discovered a metal object. Now this was applied to the Presidential body. The country waited through eleven anguishing weeks as Bell explored in vain for the bullet and as various other practitioners tried to bring relief with their scientific systems. None succeeded—though perhaps Bell would have found it if the bedstead itself had not been metal.

As the century changed and Bell and his wife aged, they spent increasing amounts of time each year at an estate built on the slopes above the Bras d'Or Lakes in Nova Scotia's Cape Breton Island. There he worked, as ever, through the night hours, slept as late as he was allowed to each morning, played with grandchildren, and experimented with large-scaled ideas. These included not only the famous, often-pictured tetrahedral kite but also a series of dramatic, fast-flying airplanes and hydroplane boats, which appeared to be heralds of the new century.

In 1891, Bell was asked to address a meeting of inventors from around the world. He told them: " … wherever you find the inventor, you may give him wealth or you may take from him all that he has; and he will go on inventing. He can no more help inventing than he can help thinking or breathing." Thus it was with this unusual figure until the time of his death in 1922. His wife, desolate, survived him by only five months.

Although Alexander Graham Bell may have wandered apart from the usual path of the American inventor, his influence on subsequent generations of inventors was impressive. This is not only because of the quality of his imaginings—the drawings for his tetrahedral kites, for example, reveal a mind virtually snapping with ideas—but also because of his individualistic approach to invention: He looked into what interested him. This ex-Scotsman who loved the highlands was no scout dog for industrialists or capitalists of America's new century. This was a free and humane figure. Of course he created a whole new branch of the communications industry and lived a grand life. But he also fulfilled his own given mission of helping the deaf to communicate—and to believe in their own capacities.

*Communication is the essence of this photo of
Bell flying a kite with his friend Helen Keller on
Cape Breton Island in 1901.*

It is no surprise, therefore, that Helen Keller, who adored him, dedicated her first book to him, in these words

To
Alexander Graham Bell
Who has taught the deaf to speak
and enabled the listening ear to hear
speech from the Atlantic to the Rockies

I Dedicate the Story
of My Life

At left is Alexander Graham Bell shortly before his death in 1922.

"DAMN'D IF THEY AIN'T FLEW!"— THE WRIGHT BROTHERS

Bishop Milton Wright of the United Brethren Church taught his five children, including sons Wilbur and Orville, that the world outside their frame house in Dayton, Ohio, was a cruel and devilish place. They should be on their guard, play their cards close to the vest. And on their guard they were, never marrying, always wearing the stiff collars and button-top shoes of their old-time, small-city culture. Yet they taught the world to fly and, in doing so, gave mankind a new-century way of looking at the world. In the words of European historian Charles Dollfus, "they changed the face of the globe." And, in a strange way that puzzles historians to this day, these seemingly naive, very parochial brothers knew exactly what they were doing when it came to the business of globe-spanning inventions.

Their accomplishments in the history of aviation quite rightly struck President Taft as a masterpiece of American determination. Although the Wright brothers were devoted to the abstract idea of flight in much the same, pure way that Alexander Graham Bell was dedicated to the science of acoustics, they were Main Street businessmen in a completely different way from Bell. The Wright Cycle Company, originally located at 1127 West Third Street in Dayton, now may be seen, reconstructed, in Henry Ford's Dearborn Village. The dusty, cluttered shop dramatizes the message that America's most brilliant tinkers, while influencing the way the world would turn, and while building great fortunes, remained the same ledger-conscious young men who had switched off their own office lights and walked home late at night worrying about how to do better tomorrow.

These twin portraits of Wilbur (below) and Orville (above) Wright hint at the contrast of the two—Wilbur older, brainier, and shyer; Orville younger, riskier, and more outgoing.

Wilbur and Orville Wright, in the view of Mitchell Wilson, were the last of the American individualists—that breed of home-grown inventor who believes he not only can create a superior rat trap but also can sell it successfully to the kings of Europe and the potentates of Asia. This belief they shared with Fulton and Morse and dozens of others who had gone before them. Yet they were at the same time, in their own eyes, scientists. Their ceaseless experiments conducted with crude but carefully controlled equipment and their exhaustive reading of professional papers certainly entitled them to join the company of Sir Francis Bacon's heirs, even though their formal education never went beyond high school.

One example of their probing minds (and their intention of contributing to science) is given by the key letter that Wilbur wrote to the Smithsonian Institution in the last year of the old century. Then aged thirty-two, he introduced himself by saying he had been interested in flight since boyhood, and explained:

> I am about to begin a systematic study of the subject in preparation for practical work to which I expect to devote what time I can spare from my regular business. I wish to obtain such papers as the Smithsonian Institution has published on this subject, and if possible a list of other works in print in the English language … I wish to avail myself of all that is already known and then if possible add my mite to help on the future worker who will attain final success.

Impressed, the Smithsonian archivists sent along all the information they could supply, which was considerable. And it was on the basis of this research that Wilbur, joined by his younger brother and working partner Orville, began the methodical analysis of how to make "aeroplanes"—by which he meant wings that could supply lift to heavier-than-air vehicles.

THE
American Star Bicycle.

———o———

PATENTED BY G. W. PRESSEY.

AMERICA WHEELS TO THE FRONT.

H. B. SMITH MACHINE CO.   SMITHVILLE. N. J.

The above engraving represents the *New Bicycle* with all the recently added improvements; among which may be mentioned the *framing* which is light and strong, and less liable to be broken in case of accident, than the expensive back-bone usually employed in the construction of crank machines. Also the improved arrangement of the propelling treadles, by which the friction on the main bearing is reduced to a minimum, thus dispensing with the necessity of "Ball Bearings" and other perishable supports.

The *Brake* is of improved construction and may be applied at will without removing the hands from the steering handles. Any degree of friction may be applied even sufficient to lock the carrying wheel, and this too without any possible danger of a "Header."

The construction of the *Wheel* has also been improved, retaining that valuable feature of tangency and making the spokes "Direct Acting," the ends being *upset* large enough so that the thread will not weaken them, and on the outer ends heads are formed with a *die*, thus insuring a perfect rivet head. The spokes are screwed into the hub and being adjustable the rider can always keep his wheel true, or easily replace a broken one should he be so unfortunate as to knock one out.

The *Saddle* is of the finest "Suspension" style, and made by ourselves. The frame is of steel and the covering is genuine Pig Skin. The saddle is placed on a long wide spring which covers the wheel, and may be placed either forward or back to suit the rider; the whole being most comfortable to sit and ride upon.

For prices and further particulars see next page.

When the Wright Brothers were experimenting with bicycles, this dashing rival appeared on the market. Though bizarre, it had the right motto for the time: America Wheels to the Front.

The young men had teamed up to work together ever since the time when Orville, a teenager disinterested in school, had opened shop as a printer. He had given himself the impossible-sounding assignment of building a large-sized press for his own, local newspaper. By scrounging and making-do, he had succeeded in producing a workable machine. Intrigued, Wilbur (who was recovering from the long-term, depressive effects of an athletic accident) joined him as editor. Though at first moderately successful, the press and its products were soon challenged by papers produced in nearby big cities—imitators of Whitelaw Reid's and William Randolph Hearst's mass-circulation journals. Closing their shop but vowing to go together into another business that was on the rise, the brothers opened their bicycle shop. They would figure out how to build the best bikes for the money.

It was not that the brothers never fought or argued; they did this late into the night, to the annoyance of other, sleep-starved family members. They debated, for example, the principles on which a toy that their father had given them worked—a helicopter, an inspired gift of unimaginable importance. As their creative discussions about things mechanical progressed night and day, the imagined objects became fully dimensional, assumed new and constructible shapes. Biographer J. G. Crowther expressed the process and the men this way:

> The unique bond of intellectual collaboration was not based on similarity of temperament, but rather on contrast. They were complementary to each other. Apart from their passionate interest in the same thing, almost the only characteristic they had in common was grey-blue eyes … .The extraordinary collaboration between the brothers, unparalleled in the history of science, seemed to some extent to be a repetition in the realm of intellectual understanding of the deep affection that had united their father and mother.

Whatever the psychological explanation, the combined brothers constituted a dynamic force which succeeded wondrously in the field of aeronautics at a time when many of the world's leading engineers (including Bell and Edison) were pondering the same question of how to give man wings. It was a time when Edison seemed to have proved that inventions could be delivered, on order. If man desired to fly in the new century—as he had desired to banish darkness, to send messages around the world, to travel across the continents in luxury—then surely in this new century that wish too could be granted by the demigods of invention.

In their humble bicycle shop the Wright brothers soon learned the crucial fact that their pricey "safety bike" and other go-fast products could be sold in great numbers *because Americans loved to be rapidly propelled,* even at considerable expense. They felt certain that if they could produce a similar but even more complex technological structure, a flying machine powered by an automobile-type engine, it would find a market. That conviction, implied but never expressed in their writings, was as strong a motivation as their very real fascination with the theories of flight. To say it again: These skinny, gray-blue-eyed, self-educated, gently mannered brothers knew exactly what they were doing and why. As inventors and businessmen with a presumably superior and innovative product for a large market, they had the right stuff to play a major role in the century that lay just around the corner.

Most important among the papers that Orville Wright received in 1899 from the Smithsonian were reports of the glider experiments carried out by a German scientist named Otto Lilienthal. In a number of remarkable glides (more than two thousand glides, in fact) he demonstrated and measured with great precision the advantage of curved wings over flat wings for helping sustain flight. Along with that gift to the Wright brothers and to other inquirers into the aeronautical facts of life, Lilienthal bequeathed the concept of equilibrium—the idea that a body in flight must be able to balance itself on the wind just as bicyclists balance themselves between earth and air.

Lilienthal accompanied his breathtaking, well-photographed flights with intensive studies of birds and how they fly. His most valuable book was entitled *Bird Flight as Basis of the Art of Flying* (*Der Vogelflug als Grundlage der Fliegekunst,* 1889). To the dismay of the aeronautical world, Lilienthal's ultimate flying machine was upset by a sudden gust of wind near Rhinow; he perished in the crash.

"No bird soars in a calm," Wilbur Wright reminded himself and his brother as they concluded their studies of Lilienthal's works and prepared to build a series of experimental kites. This was the realization that prompted them to get in touch with meteorological authorities: Where could they be assured of constant winds to help their painstakingly constructed kites soar into the heavens? The answer came back swiftly, and in the fall of 1900 the brothers found themselves heading for the Outer Banks of North Carolina, whose sand dunes moan to the tune of Atlantic breezes throughout the year.

In addition to the lifting breeze, the brothers recognized that they needed a system of controls to keep their high-flying kites in trim, in equilibrium. Having studied and restudied how racers control their bicycles on turns, they concluded that control in midair was a three-dimensional problem. The kite, if not properly controlled, might roll (meaning to tip to one side or another from the fore-aft axis) or it might pitch (meaning to nose-dive upon the horizontal axis) or it might yaw (a stomach-wrenching combination of the two other movements). To control the pitch, they designed a so-called "elevator" that rode out in front of the biplane kite. Rolling and yawing were controlled by lines that "warped" the wings—the way birds change the shape and raise or lower the tips of their wings.

For their 1900 kite, a magnificent, delicate structure whose wings spanned seventeen feet (total wing area, 165 square feet), the control lines were led down to the ground. Successful flights followed one after another, even as heavier weights were added to simulate a man aboard. On a few, wild occasions the brothers risked manned flights, with the pilot handling the controls.

Encouraged by their kite's well-behaved performance, they returned to Kitty Hawk, North Carolina, the next year with a larger version. But this one (the largest ever flown, with wings of twenty-two feet) seemed far less cooperative. Furthermore, the larger amount of wing area, 290 square feet, failed to provide the lift that formulas derived from Lilienthal's figures had forecast. Would it ever be possible to lift both a man and an engine, they wondered? With the exact thoroughness of laboratory scientists, they then constructed back home in their bicycle shop a wind tunnel for testing the effectiveness of variously shaped wings—discovering to their joy and excitement that, though the Lilienthal data were indeed misleading, they could be corrected. It also appeared that variable forces—namely lift and drag—could be adjusted in such a way that the wings might carry an engine-driven vehicle on the wind.

Photo at right above shows Orville Wright and Lt. Thomas Selfridge at the controls of a Wright Flyer, preparing to take off for a test at the Army's Fort Myer, VA, field in 1908. Photo below shows the airplane just after it struck the ground, killing Selfridge and badly injuring Orville.

The rest of the story is well known: the splendid behavior of the 1902 kite (particularly after a moveable rudder had been added), the building of the 1903 "Wright Flyer" as well as the design of efficient propellers, and the successful launch and flight of December 17, 1903. The power plant which provided the thrust (and which turned this plane into a vehicle quite different from a wind-controlled glider) was a four-cylinder, 179-pound gasoline engine of the Wrights' own design and construction. It produced twelve horsepower at

The next year, 1909, a Wright Flyer successfully negotiated the course at Fort Meyer (left), leading the Signal Corps to adopt the plane for Army use—a key event in the Wrights' fortunes.

1,090 rpm, roaring deafeningly in the ears of pilot and handlers alike. Above that noise, one of the five local helpers, a coastal lifesaver, hollered at the time of Orville's 120-foot, twelve-second flight: "They've done it! They've done it! Damn'd if they ain't flew!"

With winter taking charge of the Carolina dunes, the brothers retreated to Dayton, knowing that they had accomplished something totally unique. Before leaving they telegraphed the bishop that they had succeeded (there had actually been a total of four successful flights on that historic day, the longest of which was for 852 feet), that they would be home for Christmas, and that the previously worked out plan for handling press relations should be maintained. They immediately went to work on the construction of a commercially marketable aircraft—while the world contemplated the news (garbled and inaccurately repeated) of the first, historic flight. Editorialists wondered whether anything particularly useful had been achieved. Fly—for what? And who were these guys from Dayton, anyway?

The skepticism with which Americans and Europeans greeted reports of the Wright brothers' brief flights above the dunes may be understandable. The year 1903 was also when one of the most distinguished scientists in the world and one of the most vigorous proponents of manned flight suffered a massive indignity. This was astronomer Samuel Pierpont Langley, then Secretary of the Smithsonian, who had been building and flying models of powered aircraft since 1896. Backed by government funds and cheered on by his friend Alexander Graham Bell, he staged a highly publicized "aerodrome" launching from a houseboat in the Potomac River virtually simultaneously with the Wrights' flight in North Carolina.

Langley thought he understood better than anyone else the principles necessary for flying: lots of power, cambered wings, and a well-paid mechanic/pilot. He cared little about all this talk of "equilibrium." His great machine, powered by a fifty-two horsepower engine and launched into the air by a massive catapult, immediately assumed the flight pattern of a dropped stone and splashed into the river to its total destruction.

The highly amused *Washington Post* called Langley's aerodrome "the Buzzard"; editors of the *Chicago Tribune* pontificated that they had known all along that God intended mortal man to remain earthbound until the rolls were called in heaven. *Puck* magazine, the ultimate in worldwide sophistication, gibed that inventors of airplanes would surely succeed, just as soon as the law of gravity could be repealed. This climate of opinion was largely responsible for the odd-seeming fact that no one in the United States—not even the Army, which had expressed some mild interest in airplanes for observation and scouting purposes but certainly not for fighting—paid serious attention to the Wrights' invention for six wasted years.

It was not until August 2, 1909, that the Army's Signal Corps, after staging a year-long test, accepted the first "Wright Flyer" for military use—the first airplane purchased and placed in service by any government. During the test of the preceding year, the Flyer, piloted by Orville, went out of control at one point because of a broken propeller; passenger Lieutenant Thomas Selfridge (who had worked, earlier, with Bell) suffered a skull fracture and died. Despite that crash (which Orville survived, despite severe injuries), the wood-and-canvas airplanes produced at the Wright factory in Dayton soon won much-deserved respect for their careful, if violinlike, construction and for their dependability.

Undoubtedly the reason for the Wright brothers' breakthrough into acceptability as inventors and manufacturers of an aircraft capable of carrying the nation into the new century was the grand reception given the Flyers when the brothers took their act

Subtext to this prideful cartoon in the Cleveland leader might read: "President Taft and the rest of America followed far behind European princes in recognizing the Wright Brothers."

The Boy—"Gee! I'd rather be Wright than be President."

- From the Cleveland Leader

to France and Germany. Crowds by the thousands hailed the Wright *frères* and the *Gebrüder* Wright as heroes of the future; lads in France were proud to show off to friends their "Veelbur Reet" caps. Acquaintance with the brothers was sought by Edward II of England; King Alfonso of Spain took an on-the-ground flying lesson from Wilbur. At international airshows—commencing with the opener at Rheims in 1909—Wright planes proved to the world that safe flights well over a hundred feet high and many miles long were now routine. And always, after dazzling displays of figure eights and perfect circles, the planes landed gently and precisely, the controls functioning perfectly.

The brothers returned to Dayton in June of 1909 for a tumultuous, bunting-draped homecoming parade. Accepted by the world, the young prophets could now be hailed at home—America's natural heroes. Then, in the fall of that year, New York City politicos invited the brothers to be the main attraction in a stupendous event: the 300th anniversary of Henry Hudson's landing and the 100th anniversary of Fulton's eternally popular steamboat voyage. Wilbur, knowing full well the value of such a public demonstration, outdid expectations by flying twenty-one miles upriver from Governor's Island to Grant's tomb and back. Knowing also that much of the flight would be over water, he installed a watertight canoe below the Flyer's lower wing in case of emergency. The crowds loved the daring, the show. When might they themselves fly?

The time had come for the Wright brothers to put the next phase of their plan into effect: a manufacturing and sales corporation. Back in 1903 their application for a U.S. patent had been turned down (perhaps it had not been legalistic enough, they surmised). Hiring themselves a brilliant attorney named Harry A. Toulmin, they then fought the matter through the bureaucratic processes successfully, with Toulmin finally getting the brothers to agree that what they were patenting was not a flying machine per se but a three-axis system of equilibrium. The patent, the tremendously important patent, was finally granted on May 23, 1906. With that in hand, and with enough money coming in from contracts arranged with European distributors, they prepared to make a place alongside Edison among the corporate giants; they too would be engineers of the way twentieth-century America would live.

When in New York for the Hudson-Fulton show, a bright young man named Clinton R. Peterkin had approached Wilbur, claiming that his contacts with J. P. Morgan enabled him to make the contacts which the Wrights would need for funding their corporation. Wilbur, without too much confidence, let him go ahead. And, to his surprise (given the cruel and devilish nature of the secular world), Peterkin came through—with Morgan and a host of other backers, including Cornelius Vanderbilt, August Belmont, and Howard Gould. The Wright Company, with offices on Fifth Avenue, was incorporated on November 22, 1909, with a capital stock of $1 million. Wilbur became president, Orville vice-president; between them they held one-third of the stock. For each plane turned out by their Dayton plant, they would receive a ten percent royalty.

Their careful plans now seemed near fulfillment. As their lawyers successfully fought off repeated attempts at patent infringement, and as more improved, solid planes were designed and built for all the world's skies, their income increased to a point well beyond mere comfort. The brothers found themselves discussing a family dream house.

This would, in fact, become an impressive structure named Hawthorne Hill on a sev-enteen-acre plot outside of Dayton. It looked like the ultimate rich man's mansion, yet it represented to them not wealth or ostentation but family accomplishment. In an early stage of the design, Wilbur had admonished Orville that he had allocated too much space for hallways and not enough for bedrooms. "In any event," Wilbur wrote from Europe in the spring of 1911, "I am going to have a bathroom of my own, so please make me one."

On Thursday, May 2 of that year, the family made a symbolic trip up the slope on which Hawthorne Hill would be built. The brothers led their father along between them; they were accompanied by their sister Katherine, who had served as the female presence in their lives ever since the death of their mother in 1889. She had also ac-companied them on their triumphant European sorties, had flown in the Flyer, skirts tied tightly around her ankles with dependable twine. They all agreed, this hillside would be the ideal place for them to reside, to stamp the seal of their family covenant; four fluted columns should support the entrance portico.

Sadly, Wilbur never walked through those columns. He died of typhoid (possibly brought on by shellfish-poisoning) on May 30, 1912, aged 45 years. How strange that was the very same death age of Samuel Finley Breese Morse and Samuel Colt. Was this the typical fate of inventors of that day? Or, possibly, was Wilbur, never strong as a result of his youthful athletic accident and punished by years of stress and overwork, simply not healthy enough to withstand a severe and not necessarily fatal disease?

Whatever the explanation for Wilbur's early death, Orville was devastated. That, combined with a ruinous flood and fire in Dayton the next year and his father's death in 1916, pushed Orville further into himself. He retreated to his old office in the bicycle shop, rather than assuming the large chair at the airplane factory to which his new presidency entitled him. There he followed a course of letting letters go unan-swered and leaving corporate decisions unmade.

He and his sister now lived alone in Hawthorne Hill, Wilbur's bedroom left con-spicuously empty. Not long after World War I, an event occurred in Oberlin, Ohio (where Katherine had gone to college), which was even more difficult for Orville to understand than any family death. Katherine, aged fifty-three, married a classmate and moved out of the house. Friends described Orville's mood as "alternately furious and inconsolable"; Katherine had broken the family covenant. Unyieldingly, he would have nothing more to do with her. Finally, at the time of her death in 1929, Orville ap-peared at her bedside, quietly weeping.

Orville had retired from piloting airplanes a few years after Wilbur's death, yet he tried to contribute to the further development of American aviation. And he lived to play an advisory role to the government in both world wars. But, with enclosed cock-pits changing the nature of airplane construction, and with the sale of the Wright Company to a financial consortium, Orville found himself pretty much out of busi-ness—which is precisely where he wanted to be.

Orville Wright died of a heart attack in a Dayton hospital in 1948, aged seventy-seven—though perhaps it could be said he had died much earlier. The careful plans that he and Wilbur had enjoyed making together had certainly not anticipated any-thing like the jet age, or the kind of high-flying culture that the twentieth century would bring. In that corporate culture of big cities with international airports, where were the individualists, the small-town values, the tight family relationships? Modern America seemed a kind of accident, an accidental development quite different from the normal world they had expected to endure.

GEORGE EASTMAN'S MAGIC BLACK BOX

What George Eastman contributed to America as the century turned was, by contrast, no accident at all. Neither was his death; it was all quite calculated and corporate.

Whereas the Wright brothers may be seen as the final flower in the line of individualistic American inventors, George Eastman—who founded the Eastman Kodak Company in 1892—marks the arrival of that new species, the American corporate inventor. Many would follow in his wake, including Leo Baekland (associate of Eastman and founder of the Bakelite Corporation) and Lee De Forest (pioneer of radio and television) and Wallace Carrothers (who perfected long-chain polymers for the Du Pont Corporation). Yet, though the day of the Yankee tinkerer and the "happy hooligan" inventor appeared to be over, that spirit would linger forever within the evolved, more scientific, more corporate man. The relationship between the two conflicting strains can be discerned in Eastman's life.

They said of him that he was constantly "at war with himself." And of that war came great creativity, as well as ultimate destruction. Born in 1854 to a Rochester, New York, family of New England lineage, he had to face the fact as a boy that his father had died early, after years of fruitless struggle. George was determined not only to support his mother but also to show the business world that he could become not the slave of it but the master.

Dutifully, George took successive jobs in local insurance companies and banks, impressing all by his diligence and obedience. His one deviation from the role of perfect clerk was a tendency to tinker at his workbench at home through the night—that and a passion for the new field of photography. By 1877 he was taking courses in photography, learning about the painfully exacting, "wet collodion" process. He also learned of Morse's improvements on Daguerre's systems, understanding, thereby, that photography was hardly a stabilized art.

By his mid-twenties, George Eastman considered himself a competent photographer but still an amateur. He liked to tell the story of how, on a trip West, he met a group of tourists who, seeing him set up his black tent and full array of chemical equipment, assumed he was a professional portraitist and organized themselves for the inevitable group shot. They were infuriated when he declined to take their picture on grounds of amateur status. How in the world, they asked themselves, could anyone go to all that trouble, carrying around that weighty, bulky field laboratory, and not be a pro?

Experienced in the traditional modes of photography, Eastman sought ways to simplify them. His first idea was to manufacture dry plates and to market this simpler system among other photographers. To sharpen that and other burgeoning ideas, he sailed off to London, then photographic capital of the world. He returned with a plan firmly in place; on July 22, 1879, he received a patent for the new plates. He intended to produce and sell them on a large scale, knowing full well that, in America, the swiftest way to an industrial fortune was through mass production.

Backing young Eastman was a sometime boarder at his mother's house, a certain Colonel Henry Alvah Strong (who, by that and other timely investments, became a

In this portrait of George Eastman, he looks precisely like what he had been before his inventions succeeded—a small-city clerk. His Kodak Primer, which took pictures by a pull of the string and advanced film by a turnkey on top, appears below.

wealthy and notable philanthropist). On the strength of Strong's first $1,000, Eastman put himself in business, insisting on supplying only tested and foolproof products, at the lowest possible price. He believed that, beyond those well-tested American marketing principles, he needed to undertake extensive advertising (along the lines of Cyrus McCormick and his instantly well-known reaper). He would also need to establish foreign distribution—for that was another basic principle of the new American marketing, recognized as such by Morse and the Wright brothers in their turns.

By the end of his first year in business, Strong had invested another $4,000 in the enterprise; Eastman himself matched those dollars. The prospects looked bright, with sales of 4,000 units a month, until, suddenly, word from sales offices came back that the emulsion on the plates was imperfect—the gelatin in the mixture had been faulty, Eastman learned later. Working tirelessly to correct the flaw, Eastman pitched a hammock in the laboratory so he would be ever closer to the work; his meals were also cooked right there, where he worked. After too long away from home, his mother came and hauled him back by the ear to bed and bath.

Eventually Eastman hired a chemistry professor from the University of Rochester to conduct further experiments. But, even after the problem of the emulsion was overcome, Eastman remained dissatisfied with the product; what he wanted most of all was a truly simple system that would make every American tourist a potential customer, a happy amateur photographer. Ultimately, as he thought this through, he concluded in 1884 that the right way to look at his business was not as a line of products but as a total system. Rather than supplying plates—or his new invention, "film"—and cameras and lenses, he would offer his customers a simple black box costing $25. After taking their shots, customers would return the magic box to Rochester for photo development; for another $10, additional film would be inserted.

This system was soon patented and perfected (by 1891, the cameras themselves no longer had to be returned to Rochester for servicing). The magic box weighed only twenty-two ounces and was virtually foolproof. Photographic houses in New York and other cities desperately fought this simplified system; Eastman replied with even more aggressive campaigns. He had always believed the letter K to be attractive, two K's as even better; hence his invention of the name Kodak. Along with it went the tremendously appealing phrase: "You press the button—we'll do the rest."

Eastman launched this idea, this totally new industry upon the world at precisely the right time. In the Gay Nineties, Americans and Europeans were fascinated with themselves and their newly discovered powers. They wanted to be memorialized, in all their finery and autos and even, later, airplanes. They would buy this marvelous device as if it were so many pieces of bread, essential for life, even though it had never before existed in the history of the human race.

George Eastman, who collaborated with Edison in the production of film for movies, continued to stay at the head of his very own industry, unrivaled and triumphant. He saw to it that his factories were run on a profit-sharing basis; his philanthropic contributions would exceeded $100 million (the principal beneficiaries being the University of Rochester, the Eastman School of Music, and MIT, as well as Tuskegee and The Hampton Institutes). He lived to be seventy-eight, a long and tremendously fruitful life, even though the creative, private, and sensitive individual may have continued to war with the superbly organized, corporate, and public genius. He remained unmarried, alone even in the midst of his busy city and world of the new century. Stating that "My work is done … why wait?" he shot himself in 1932. The event was generally interpreted more as another smart decision than as a tragic event.

Among his many other brilliant contributions to the new century was the reborn idea that the individual inventor, though corporation-financed and -directed, could once again work for the sake of a social purpose. Like Winthrop and Franklin and Lowell, Eastman respected the ideal of the community—even when the community expanded to become the nation and the world. By that visionary social perspective as well as by the nature of his inventions, he may be judged a beneficial factor in the shaping of the modern world.

Unlike Fulton and Colt, George Eastman made neither explosives nor great profits at the expense of warring factions. Instead, he gave Americans a greater opportunity to understand themselves and to pursue happiness, camera in hand.

In the critical years between 1870 and 1920, the major institutions of our society were altered: family life, private property, the church, and the university, as well as the modes of travel and the ways of conducting business. Certain inventors, for good or ill, had been high-stake agents in that change, playing the game as if their lives (and ours) depended on it. They did.

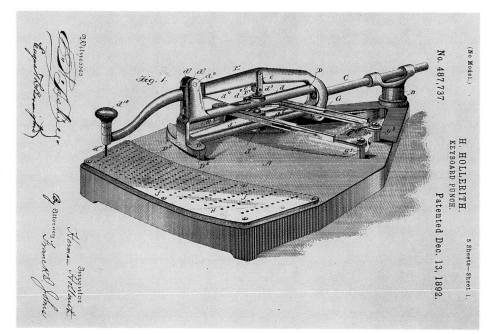

Herman Hollerith's invention of a keyboard punch for recording data, pictured in the patent application at right, anticipated the computer—instrumental in today's communications corporations.

BIBLIOGRAPHY

Wealth and fame are coveted by all men; but
the hope of wealth or the desire for fame
will never make an inventor … . You may
give him wealth or you may take away from
him all that he has; and he will go on inventing.
He can no more help inventing than he can help
thinking or breathing. Inventors are born not made.

—Alexander Graham Bell, as quoted in
Robert V. Bruce's *Conquest of Solitude*

I. GENERAL WORKS

Boorstin, Daniel J. *The Americans: the National Experience.* New York: Vintage Books, 1965.

Bruno, Leonard. *The Tradition of Technology.* Washington, D.C.: Library of Congress, 1995.

Burns, James McGregor. *The Vineyard of Liberty.* New York: Knopf, 1982.

Franklin, Benjamin. *The Autobiography of Benjamin Franklin.* New York: Pocket Books, 1939.

Hawke, David F. *Nuts and Bolts of the Past: A History of American Technology, 1776-1860.* New York: Harper, 1988.

Hindle, Brooke and Lubar, Stephen. *Engines of Change: The American Industrial Revolution, 1790-1860.* Washington, D.C.: Smithsonian Press, 1986.

Kaempffert, Waldemar (ed.). *A Popular History of American Invention.* New York: AMS Press, 1924.

Kammen, Michael. *Mystic Chords of Memory.* New York: Knopf, 1991.

Kennedy, Roger G. *Architecture, Men, Women, and Money in America, 1600-1860.* New York: Random House, 1985.

MacDonald, Anne L. *Feminine Ingenuity: Women and Invention in America.* New York: Ballentine, 1992.

Miller, John C. *Alexander Hamilton and the Growth of the New Nation.* New York: Harper, 1959.

Morison, Samuel Eliot. *The Oxford History of the American People.* New York: Oxford University Press, 1965.

Peterson, Merrill. *Thomas Jefferson and the New Nation.* New York: Oxford University Press, 1970.

Smith, Page. *The Shaping of America: A People's History of the Young Republic.* New York: McGraw-Hill, 1980.

Stein, Ralph. *The Great Inventions.* New York: Simon & Schuster, 1976.

Strandh, Sigvard. *A History of the Machine.* New York: A&W Publishsing, 1979.

Tocqueville, Alexis de. *Democracy in America* (2 vols). New York: Knopf, 1948.

Weber, Robert J. Forks, *Phonographs and Hot Air Balloons: A Field Guide to Inventive Thinking.* New York: Oxford University Press.

Wilson, Mitchell. *American Science and Invention.* New York: Simon & Schuster, 1954.

II. CHAPTER ONE

Adams, James Truslow. *Revolutionary New England, 1691-1776.* Boston: Atlantic Monthly, 1923.

Albion, Robert G. *New England and the Sea.* Mystic: Mystic Seaport Museum, 1972.

Barker, Shirley. *Builders of New England.* New York: Dodd, Mead, 1965.

Beals, Careleton. *Our Yankee Heritage: New England's Contribution to American Civilization.* New York: McKay, 1955.

Benes, Peter (ed.). *The Bay and the River, 1600-1900.* Boston: Dublin Seminars, 1982.

Black, Robert C., III. *The Younger John Winthrop.* New York: Columbia University Press, 1966.

Bruchy, Stuart (ed.). *The Colonial Merchant, Sources and Readings.* New York: Harcourt, 1966.

Greenberg, Dolores. "Energy, Power, and Perceptions of Social Change in the Early Nineteenth Century." The American Historical Review, June 1990.

Hindle, Brooke (ed.). *America's Wooden Age: Aspects of its Early Technology.* Tarrytown, NY: Sleepy Hollow Press, 1975.

Morison, Samuel Eliot. *Builders of the Bay Colony.* Boston: Houghton Mifflin, 1930.

Powell, Sumner Chilton. *Puritan Village: The Formation of a New England Town.* Middletown, CT: Wesleyan University Press, 1963.

Sloan, Eric. *A Museum of Early American Tools.* New York: Ballentine, 1964.

Stapleton, Darwin H. *The Transfer of Early Industrial Technology to America.* Philadelphia: American Philosophical Society, 1987.

Tager, Jack and Wilkie, Richard W. *Historical Atlas of Massachusetts.* Amherst, MA: University of Massachusetts Press, 1991.

III. CHAPTER TWO

Adams, Russell B., Jr. *The Boston Money Tree.* New York: Crowell, 1977.

Asher, Robert. *Connecticut Workers and Technological Change.* Storrs, CT: University of Connecticut Press, 1983.

Bailey, Chris H. *Two Hundred Years of American Clocks and Watches.* Englewood Cliffs, NJ: Prentice Hall, 1975.

Billington, James H. *Fire in the Minds of Men: Origins of the Revolutionary Faith.* New York: Basic Books, 1980.

Carter, Edward C., III. *Benjamin Latrobe and Public Work: Professionalism, Private Interest, and Public Policy in the Age of Jefferson.* Washington, D.C.: Public Works Historical Society, 1976.

Dalzell, Robert F. *Enterprizing Elite: the Boston Associates and the World They Made.* Cambridge, MA: Harvard University Press, 1987.

Dangerfield, George. *The Awakening of American Nationalism, 1815-1828.* New York: Harper, 1965.

Deetz, James F. *In Small Things Forgotten: The Archaeology of Early American Life.* Garden City, NY: Doubleday, 1977.

Ferguson, Eugene S. *Oliver Evans: Inventive Genius of the American Industrial Revolution.* Greenville, DE: Hagley, 1980.

Grant, Ellsworth S. *Yankee Dreamers and Doers.* Chester, CT: Pequot Press, 1973.

Kulik, Gary (ed.). *The New England Mill Village, 1790-1860.* Cambridge: MIT Press, 1982

Larkin, Jack. *The Reshaping of Everyday Life, 1790-1840.* New York: Harper, 1988.

McDonald, Forrest. *Alexander Hamilton, A Biography.* New York: W.W. Norton, 1979.

Prude, Jonathan. *The Coming of Industrial Order: Town and Factory Life in Rural Massachusetts.* New York: Cambridge University Press, 1983.

Schlesinger, *Arthur M., Jr. The Age of Jackson.* Boston: Little, Brown, 1946.

Smith, Page. *The Nation Comes of Age: A People's History of the Ante-Bellum Years.* New York: McGraw-Hill, 1981.

IV. CHAPTER THREE

Ackerman, Carl W. *George Eastman.* Boston: Houghton Mifflin, 1930.

Adams, James Truslow. *New England in the Republic, 1776-1850.* Boston: Little, Brown, 1926.

Bruce, Robert V. *Alexander Graham Bell and the Conquest of Solitude.* Ithaca: Cornell University Press, 1973.

Combs, Harry. *Kill Devil Hill: Discovering the Secret of the Wright Brothers.* Boston: Houghton Mifflin, 1979.

Conot, Robert. *A Streak of Luck.* New York: Seaview Books, 1979.

Crouch, Tom. *The Bishop's Boys: A Life of Wilbur and Orville Wright.* New York: W.W. Norton, 1989.

Grant, Ellsworth S. *The Colt Legacy: The Colt Armory in Hartford, 1855-1980.* Providence: Mowbray, 1982.

Halion, Richard P. (ed.). *The Wright Brothers: Heirs of Prometheus.* Washington, D.C.: Smithsonian Press, 1978.

Howard, Fred. *Wilbur and Orville: A Biography of the Wright Brothers.* New York: Knopf, 1988.

Josephson, Matthew. *Edison.* New York: McGraw-Hill, 1959.

Lehmann-Haupt, Hellmut. *The Book in America: A History of the Making and Selling of Books in the United States.* New York: Bowker, 1952.

Mabee, Carleton. *The American Leonardo: A Life of Samuel F. B. Morse.* New York: Knopf, 1943.

Millard, Andre. *Edison and the Business of Invention.* Baltimore: The Johns Hopkins Press, 1990.

Morison, Elting, E. *Men, Machines, and Modern Times.* Cambridge, MA: MIT Press, 1966.

Newhouse, Elizabeth (ed.). *Inventors and Discoverers Changing our World.* Washington, D.C.: National Geographic Society, 1988.

Newsome, Iris (ed.). *Wonderful Inventions: Motion Pictures, Broadcasting, and Recorded Sound at the Library of Congress.* Washington, D.C.: Library of Congress, 1985.

Renstrom, Arthur G. (ed.). *Wilbur and Orville Wright, Pictorial Materials.* Washington, D.C.: Library of Congress, 1982.

Schlesinger, Carl (ed.). *The Biography of Ottmar Mergenthaler.* New Castle, DE: Oak Knoll, 1989.

SOURCES

Illustrations for this book come from various divisions within the Library of Congress. The Prints and Photographs Division (P&P) holds, among its millions of items, the following discrete collections, cites to which will be found on these pages: Historic American Buildings Survey (HABS); Historic American Engineering Record (HAER); and the Poster Collection (Pos). Illustrations for this volume also were drawn from the Rare Book and Special Collections Division (RBD); the Geography and Map Division (G&M); the Manuscript Division (MSS); the General Collections (Gen); and the Publishing Office (Publ).

Those who wish to order reproductions of illustrations in this book should contact the Library's Photoduplication Service and cite the Library of Congress negative numbers, or item call numbers provided below. *Please note:* Negative numbers begin with the following prefixes: LC-USZ62-, LC-USZ61-, LC-USZC2-(color), LC-X, LC-A1-, LC-USA7-, LC-F801-, LC-G9-, LC-USW-, LC-USZC4-(color), LC-MSS-(Manuscript Division negative), G&M neg. # (Geography and Map Division negative number). All other numbers are call numbers for the books or other materials from which the illustration was taken. When multiple images appear on a page, negative numbers will be given in the order the images are arranged, left to right, top to bottom.

Frontmatter: (opp. title page) LC-USZ62-12434; (opp. Foreword) LC-USZ62-42041; (opp. p. 1) LC-MSS-027748-71.

Chapter One: (2) LC-USZ62-2155 and LC-MSS-027748-65; (3) LC-MSS-027748-64; (4) LC-MSS-027748-107 and LC-MSS-027748-56; (5) LC-MSS-027748-24; (7) LC-USZ62-31144; (8) HAER No. NY-125-1 and HAER-NY-125, sheet 4 of 5; (9) HAER No. NY-125-8 and HAER No. NY-125-9; (10) LC-USZ62-31143 and LC-USZ62-2252; (11) HAER WV-5, sheet 3 of 4 and LC-USZA1-777; (12) LC-USZ62-37869; (14) QL737.C4C4 [RBD]; (15) SH381.C3 [RBD]; (19) LC-USZ62-3249 (image and title only) or LC-USZ62-56362 (entire page) or LC-USZC4-569 (color); (21) HAER No. NJ-1-28, NJ 16-Pat 15-28; (22) LC-USZ62-57965 and LC HABS NH, 8-Port; (27) LC-USZ62-12250; (28) LC-USZ62-43346; (29) LC-USZ62-43345 and LC-USZ62-42702; (30) LC-USZ62-44710; (31) LC-USZ62-44711.

Chapter Two: (32) LC-USZC4-256; (35) LC-USZ62-8282; (36) LC-USZ62-802; (37) LC-USZ62-37836 and LC-USZ62-37835; (38) TS2135 .U6E8 [RBD]; TS 2145 .E85, Plate VIII [Gen]; (40) TJ464 .L33, title page, [Gen] and TJ464 .L33, Plate XIII [Gen]; (41) LC-USZ62-31146; (43) LC-USZ62-12291; (44) LC-USZ62-1183; (46) LC-USA7-13145; (50) T223 .B, Feb. 27, 1872 [Gen]; (51) LC-X9-4; (53) G&M neg.

#1678; (54) LC-USZ62-24311; (55) LC-USZ62-23944; (57) LC-USZ62-41874; (62) HAER-MA-2A, sheet 1 of 3; (63) HAER MASS, 9-Low, 9B-12, HAER No. NA-2B-12; (64) HAER MASS, 9-Low 9B-18, HAER No. NA-2B-18; (65) HAER-MA-2A, sheet 3 of 3 and HAER-MA-2A, sheet 2 of 3; (66) HAER MASS, 9-Low, 8A-12, HAER No. MA-1A-12; (68) HAER MASS, 9-Low, 8-27, HAER No. MA 1-27 and HAER MASS, 9-Low, 8-30, HAER No. MA 1-30; (69) HAER MASS, 9-Low, 8A-1, HAER No. MA-1A-1; (70) HAER MASS, 9-Low, 16-2, HAER No. MA-9-2; (71) LC-USZ62-113947; (72) LC-USZ62-113946; (73) LC-USW3-16339-C and T223 .B, Dec. 10, 1846, Patent no. 4,879 [Gen]; (74) T223 .B, Feb. 27, 1872, Patent no. 124,137 [Gen] and T223 .B, Feb. 27, 1872, Patent no. 124,027 [Gen]; (75) LC-USZC4-2340 (color) or LC-USZ62-14084 and T223 .B, Feb. 27, 1872, Patent no. 124,007 [Gen].

Color Signature: (Plate 1) "Torchlight Procession Around the World," LC-USZC4-2387; (Plate 2) notes on macaroni-maker, [MSS] and "The Romance of Invention," T15 .B77 1886 [Gen] and LC-USZC4-254; (Plate 3) "Hamlin's Wizard Oil" (with elephant), LC-USZC4-1032; (Plate 4) "Wolcott's Instant Pain Annihilator," LC-USZC2-36; (Plate 5) "Old Sachem Bitters," LC-USZC4-674; (Plate 6) "Hamlin's Wizard Oil" (two people at table), P&P/Pos; (Plate 7) "The Universal Food Chopper," LC-USZC4-1154; (Plate 8) sewing machine ad, LC-USZ62-38598 (b/w) and "Home Washing Machine & Wringer," LC-USZC4-775.

Chapter Three: (76) LC-USZC4-626; (78) LC-USZC4-2341; (79) LC-USZ62-40864; (80) LC-USZ62-64180; (81) LC-USZC2-3066 (color) or LC-USZ62-101209; (85) LC-USZ62-21773; (86) LC-USZ62-114500; (87) TS535 .C6 B3 office [RBD]; (91) LC-USZ62-55184 and N325 .M6 1827, title page [RBD]; (93) LC-USZ62-664; (97) LC-USZ62-41756; (98) T223 .B, Feb. 27, 1872, Patent no. 123,984 [Gen]; (99) HAER-NJ-70-C-35; (100) LC-USZ62-8279; (101) Publ, photograph by Reid Baker; (102) LC-USZ62-13823; (103) LC-F801-19051; (104) LC-USZ62-21241; (107) LC-G9-Z4-116794-T-1; (108) LC-MSS-51268-01; (109) T223 .B, March 7, 1876, Patent no. 174,465 [Gen] and LC-G9-Z4-68812-T; (110) LC-USZ62-62787; (111) LC-G9-Z2-20,511-B and LC-G9-Z2-22,934-B; (112) LC-G9-Z1-137,772-A (both images); (114) LC-G9-Z1-137,735-A; (115) LC-USZ62-3205; (116) LC-W861-87 and LC-W861-92; (117) LC-USZ62-40822; (120) LC-USZ61-470 and LC-USZ61-473; (121) LC-USZ62-91463; (126) LC-USZ62-11170 and TR 260 .E14, 1888 [Gen]; (128) LC-USZ62-44852.

INDEX